I

Carl Wittwer   Meinhard Hahn   Karen Kaul (Eds.)

Rapid Cycle Real-Time PCR – Methods and Applications
Quantification

Springer-Verlag Berlin Heidelberg GmbH

Carl Wittwer   Meinhard Hahn   Karen Kaul  (Eds.)

# Rapid Cycle Real-Time PCR – Methods and Applications

## Quantification

With 78 Figures and 108 Tables

Springer

Professor Dr. CARL WITTWER
Director of Flow Cytometry & New Technology
Department of Pathology
University of Utah
School of Medicine
Salt Lake City, UT 84132
USA

Dr. MEINHARD HAHN
Deutsches Krebsforschungszentrum
Abt. Molekulare Genetik
Im Neuenheimer Feld 280
69120 Heidelberg
Germany

Dr. KAREN KAUL
Evanston Northwestern Healthcare
Department of Pathology
2650 Ridge Avenue
Evanston, IL 60201
USA

ISBN 978-3-540-20629-3

We thank Mr. Olfert Landt, TIB MOLBIOL Syntheselabor, Berlin, Germany, for carefully reviewing the sequence and technical information and for his valuable comments.

Cataloging-in-Publication Data applied for
Bibliographic information published by Die Deutsche Bibliothek Die Deutsche Bibliothek lists this publication in the Deutsche Nationalbibliografie; detailed bibliographic data is available in the Internet at <http://dnb.ddb.de>.
  ISBN 978-3-540-20629-3    ISBN 978-3-642-18840-4 (eBook)
  DOI 10.1007/978-3-642-18840-4

http://www.springer.de/medizin

© Springer-Verlag Berlin Heidelberg 2004
Originally published by Springer-Verlag Berlin Heidelberg New York in 2004

Cover Design: design & production GmbH, 69121 Heidelberg, Germany
Typesetting: TypoStudio Tobias Schaedla, 69120 Heidelberg, Germany

Printed on acid free paper    SPIN 10921111    18/5141 – 5 4 3 2 1 0

# Table of Contents

# Methods

# Housekeeping Genes: A Gold Standard?

Onno Bakker*, Daphne C. Timmer

## Introduction

Gene expression studies that aim at precision require normalization to an internal control or "housekeeping" gene. The greatest challenge in choosing an appropriate housekeeping gene is maintaining expression consistency during treatment.

Many genes can, in principle, be used as housekeeping genes [1]; however, the two most commonly used are glyceraldehyde-3-phophate dehydrogenase (GAPDH) and β-actin. Unfortunately, these genes are not as constant in their expression as one would think (or hope) and this particular problem has resulted in many recent publications [2–10]. β-actin shows a diurnal rhythm and the expression level changes when treating with certain hormones [4]. In addition, the presence of pseudogenes can be confounding when there is, even a little, DNA contamination in the RNA preparation [11]. Similar difficulties can occur with the well-known housekeeping genes GAPDH, elongation factor 1 alpha (EF1α), and cyclophilin [11].

Usually, an adequate housekeeping gene can be found in a controlled system (*e.g.*, cell culture) using a commercial kit or the "trial and error" method. However, more challenges arise when working with samples from patients who undergo a variety of treatments [12], because it is impossible to manipulate conditions or repeat the experiment. Similar difficulties arise when the expression of a specific gene is studied in animals throughout the day. Therefore, this chapter focuses on two specific experimental designs: (1) the expression of four different housekeeping genes studied in patient tissue samples, and (2) the expression of three housekeeping genes studied during a 12-h light and a 12-h dark cycle in rat liver.

* O. Bakker, Endocrinology & Metabolism, F5–171, Meibergdreef 9, 1105 AZ Amsterdam, The Netherlands, E-mail: o.bakker@amc.uva.nl

## Materials

Equipment    RNase-free glassware and disposables
MagNA Pure LC instrument (Roche Diagnostics, Mannheim, Germany)
LightCycler® instrument (Roche Diagnostics, Mannheim, Germany)

Reagents    Amplification primers (BioLegio, Nijmegen, The Netherlands)
High Pure Tissue mRNA kit (Roche Molecular Biochemicals, Germany)
MagNA Pure LC RNA Isolation Kit II (tissue) (Roche Molecular Biochemicals, Germany)
First Strand cDNA Synthesis Kit (Roche Molecular Biochemicals, Germany)
LightCycler – FastStart DNA Master SYBR Green I (Roche Molecular Biochemicals, Germany)
LightCycler – DNA Master SYBR Green I (Roche Molecular Biochemicals, Germany)

## Procedure

Primer Design    The primers are listed in Table 1 and have been previously published. Primer specificity was checked by comparison to the Genbank database.

Sample Preparation    Total RNA was purified from 58 human liver biopsies using the TriPure reagent. From this, cDNA was prepared using the First Strand cDNA Synthesis Kit with

**Table 1A.** Human primer sequences

| GAPDH (GenBank Accession # BC025925) | |
|---|---|
| Forward<br>Reverse | TGA ACG GGA AGC TCA CTG G (728–746)<br>TCC ACC ACC CTG TTG CTG TA (1015–1034)<br>MgCl$_2$: 4 mM |
| | Length: 306 bp |
| **EF-1α (GenBank Accession # BC057391)** | |
| Forward<br>Reverse | GAA CCA TCC AGG CCA AAT AA (1059–1078)<br>CCG TTC TTC CAC CAC TGA TT (1440–1459)<br>MgCl$_2$: 4 mM |
| | Length: 400 bp |
| **β-actin (GenBank Accession # BC004251)** | |
| Forward<br>Reverse | GGG TCA GAA GGA TTC CTA TG (202–221)<br>GGT CTC AAA CAT GAT CTG GG (420–439)<br>MgCl$_2$: 5 mM |
| | Length: 237 bp |
| **cyclophilin B (GenBank Accession # M60857)** | |
| Forward<br>Reverse | GAG ACT TCA CCA GGG G (302–317)<br>CTG TCT GTC TTG GTG CTC TCC (534–554)<br>MgCl$_2$: 4 mM |
| | Length: 252 bp |

**Table 1B.** Rat primer sequences

| GAPDH (GenBank Accession # XM214287) | | |
|---|---|---|
| Forward | AAC CAC GAG AAA TAT GAC AAC (406–426) | |
| Reverse | CAT CCT GGG CTA CAC TGA G (810–828) | |
| | MgCl$_2$: 4 mM | Length: 422 bp |
| β-actin (GenBank Accession # NM031144) | | |
| Forward | GGG TCA GAA GGA CTC CTA CG (141–160) | |
| Reverse | GGG TCT AGT ACA AAC TCT GG (359–378) | |
| | MgCl$_2$: 5 mM | Length: 237 bp |
| cyclophilin B (GenBank Accession # NM022536) | | |
| Forward | GAG ACT TCA CCA GGG G (312–327) | |
| Reverse | CTG TCC GTC TTG GTG TTC TCC (544–564) | |
| | MgCl$_2$: 4 mM | Length: 252 bp |

random primers (AMV reverse transcriptase) according to the manufacturer's protocol. The cDNA samples were a kind gift of Drs R. Peeters, T. Visser and G. van den Berghe (Rotterdam & Leuven; see reference 12)

PolyA$^+$-RNA was purified from rat liver using the MagNA Pure instrument and the mRNA tissue II kit. From this, cDNA was prepared using the First Strand cDNA Synthesis Kit with random primers (AMV reverse transcriptase).

**Table 2.** LightCycler reaction mix for a 20 µl reaction (both human and rat)

**PCR with the LightCycler**

| GAPDH / EF1α / β-actin | Volume (µl) | [Final] |
|---|---|---|
| DNA Master SYBR Green I | 2.0 | |
| Forward primer | 0.5 | 0.2 µM |
| Reverse primer | 0.5 | 0.2 µM |
| MgCl$_2$ (25 mM) | 2.4 or 3.2 | 4 or 5 mM |
| H$_2$O (sterile, PCR grade) | 12.6 or 11.8 | |

The mix was prepared for (n+1) samples. 18 µl was added to each capillary. For each sample, 2 µl of cDNA was added to each capillary.

| Cyclophilin B | Volume (µl) | [Final] |
|---|---|---|
| FastStart DNA Master SYBR Green I | 2.0 | |
| H$_2$O (sterile, PCR grade) | 12.6 | |
| Forward primer | 0.5 | 0.2 µM |
| Reverse primer | 0.5 | 0.2 µM |
| MgCl$_2$ (25 mM) | 2.4 | 4 mM |

The mix was prepared for (n+1) samples. 18 µl was added to each capillary. For each sample, 2 µl of cDNA was added to each capillary.

Sealed capillaries were spun in the LC Carousel centrifuge and placed into the LightCycler.

The LightCycler was programmed as follows:

**Table 3A.** LightCycler parameters for GAPDH and EF1α

| Program: Pre-incubation | | Type: quantification | | | Cycles: 1 |
|---|---|---|---|---|---|
| Segment Number | Temperature target (°C) | Hold Time (s) | Slope (°C/s) | Acquisition Mode | |
| 1 | 95 | 10 | 20 | None | |
| | | | | | |
| Program: Amplification | | Type: quantification | | | Cycles: 45 |
| Segment Number | Temperature target (°C) | Hold Time (s) | Slope (°C/s) | Acquisition Mode | |
| 1 | 95 | 0 | 20 | None | |
| 2 | 55 | 5 | 20 | None | |
| 3 | 72 | 10 | 20 | Single | |
| | | | | | |
| Program: Melting | Type: melting curve | Cycles: 1 | | | |
| Segment Number | Temperature target (°C) | Hold Time (s) | Slope (°C/s) | Acquisition Mode | |
| 1 | 95 | 0 | 20 | None | |
| 2 | 70 | 15 | 20 | None | |
| 3 | 95 | 0 | 0.1 | Continuous | |

**Table 3B.** LightCycler parameters for cyclophilin B

| Program: Pre-incubation | | Type: quantification | | | Cycles: 1 |
|---|---|---|---|---|---|
| Segment Number | Temperature target (°C) | Hold Time (s) | Slope (°C/s) | Acquisition Mode | |
| 1 | 95 | 600 | 20 | None | |
| | | | | | |
| Program: Amplification | | Type: quantification | | | Cycles: 45 |
| Segment Number | Temperature target (°C) | Hold Time (s) | Slope (°C/s) | Acquisition Mode | |
| 1 | 95 | 0 | 20 | None | |
| 2 | 50 | 10 | 20 | None | |
| 3 | 72 | 14 | 20 | Single | |
| | | | | | |
| Program: Melting | Type: melting curve | Cycles: 1 | | | |
| Segment Number | Temperature target (°C) | Hold Time (s) | Slope (°C/s) | Acquisition Mode | |
| 1 | 95 | 0 | 20 | None | |
| 2 | 70 | 15 | 20 | None | |
| 3 | 95 | 0 | 0.1 | Continuous | |

**Table 3C.** LightCycler parameters for β-actin

| Program: Pre-incubation | | Type: quantification | | Cycles: 1 | |
|---|---|---|---|---|---|
| Segment Number | Temperature target (°C) | Hold Time (s) | Slope (°C/s) | Acquisition Mode | |
| 1 | 95 | 10 | 20 | None | |

| Program: Amplification | | Type: quantification | | Cycles: 45 | |
|---|---|---|---|---|---|
| Segment Number | Temperature target (°C) | Hold Time (s) | Slope (°C/s) | Acquisition Mode | |
| 1 | 95 | 0 | 20 | None | |
| 2 | 50 | 5 | 20 | None | |
| 3 | 72 | 10 | 20 | Single | |

| Program: Melting | Type: melting curve | Cycles: 1 | | | |
|---|---|---|---|---|---|
| Segment Number | Temperature target (°C) | Hold Time (s) | Slope (°C/s) | Acquisition Mode | |
| 1 | 95 | 0 | 20 | None | |
| 2 | 70 | 15 | 20 | None | |
| 3 | 95 | 0 | 0.1 | Continuous | |

## Results and Discussion

In our first study, the expression of four commonly used housekeeping genes – β-actin, GAPDH, cyclophilin, and EF1α – was measured in liver biopsies of 58 patients. The mRNAs for the four genes tested were detected in all samples. The standard curve parameters were within acceptable ranges ($R^2 > 0.98$ and $-3.1 < \text{slope} < -3.6$). By using controls without reverse transcriptase, the absence of DNA was verified.

**Table 4.** p-value of Mann-Whitney U test comparing treated versus non-treated subjects

| Hormone Therapy (# treated of 58) | β-actin | cyclophilin | EF1α | GAPDH |
|---|---|---|---|---|
| Insulin (20) | 0.01 | 0.84 | 0.95 | 0.11 |
| Thyroid Hormone (21) | 0.18 | 0.05 | 0.65 | 0.07 |
| Glucocorticoids (29) | 0.40 | 0.10 | 0.21 | 0.45 |

$p < 0.05$ is considered a significant change

As shown in Table 4, housekeeping gene expression levels vary between patients according to the different treatments received. For example, β-actin expression levels are reduced due to insulin treatment while cyclophilin expression levels were modified after thyroid hormone treatment. Although of the four house-keeping genes tested, EF1α appears to be better choice than GAPDH because the higher p-values suggest less influence of hormone therapy on expression levels there were problems with this PCR which precluded its use (see chapter by Ramakers et al).

Our second study measured the expression of three different housekeeping genes in rat liver during a 12-h light and a 12-h dark cycle (circadian experiment). Many genes, including those involved in metabolic processes and common housekeeping genes, may be expressed in a cyclic fashion in different tissues [13]. Throughout the day, the expression of β-actin and cyclophilin varied up to five-fold (Figure 1). The expression of GAPDH was the least affected and would be the most appropriate housekeeping gene to use for this experimental design.

In both experimental designs studied here, it was essential to choose the correct housekeeping gene to obtain a constant expression level. If a housekeeping gene is chosen that varies with the protocol, the results of the experiment are compromised.

Instead of housekeeping genes, total RNA can be used to standardize samples by measuring them with a spectrophotometer. Unfortunately small samples (e.g., biopsies) cannot be easily quantified. Furthermore, because only 2% of total RNA is mRNA, a two-fold change in mRNA expression may not be detectable [16]. Another complication is that automated purification methods yield variable proportions of mRNA and total RNA. RNA can also be quantified with RiboGreen® dye on a fluorometer [14] or LightCycler [15]. A further option is both quantitation and quality assessment using Lab-on-a-chip™ technology (Agilent Technologies).

Vandersompele *et al.* [17] have recently published an algorithm to increase the precision of housekeeping gene expression levels. They describe a "strategy to

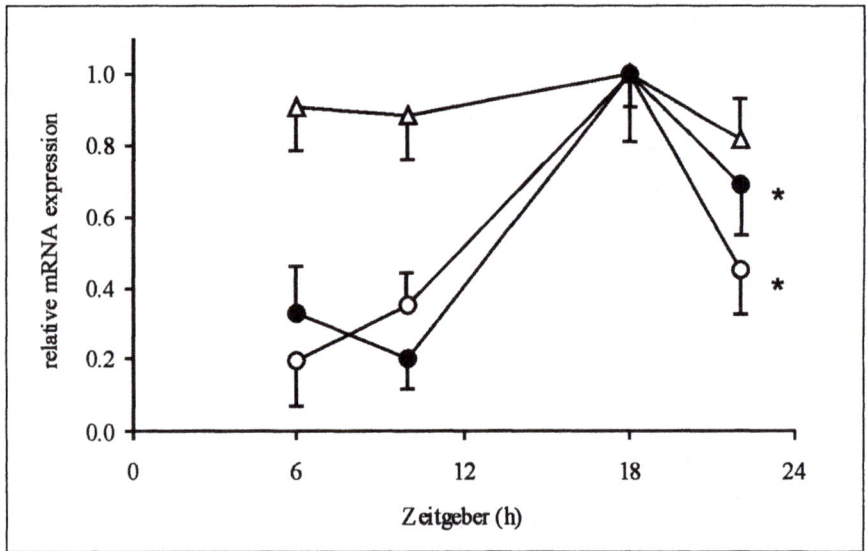

**Fig. 1.** The expression of three different housekeeping genes β-actin (white circles), GAPDH (white triangles), and cyclophilin (black circles), in rat liver during a 12-h light, 12-h dark cycle. Hour 0 is lights on (07:00 am). The points represent the mean of five individual samples; error bars indicate SEM. The β-actin and cyclophilin expression levels changed significantly throughout the day (*, $p < 0.05$, ANOVA)

identify the most stably expressed control genes in a given set of tissues, and to determine the minimum number of genes required to calculate a reliable normalization factor."

When gene expression must be normalized to compare data, it is essential to choose the appropriate housekeeping gene. Quantification of the housekeeping gene must be reproducible and validated for the particular experimental protocol. As Bustin said [2], "Today's challenge is no longer the need to develop ever more sensitive and specific quantification assays but to advance experimental protocols and designs that are rigorously controlled and allow meaningful global comparisons."

*Acknowledgements.*
A word of thanks to Drs. R. Peeters and T.J. Visser (Rotterdam, The Netherlands) and Dr. G. van den Berghe (Leuven, Belgium) for sharing their biopsy RNA samples with us. Dr. Zandieh Doulabi (Amsterdam, The Netherlands) is thanked for his help with the circadian expression experiments.

# References

1. Warrington JA, Nair A, Mahadevappa M, Tsyganskaya M. (2000) Comparison of human adult and fetal expression and identification of 535 housekeeping/maintenance genes. Physiol Genomics 2:143–147
2. Bustin SA (2002) Quantification of mRNA using real-time reverse transcription PCR (RT-PCR): trends and problems. J Mol Endocrinol 29:23–39.
3. Tricarico C, Pinzani P, Bianchi S, Paglierani M, Distante V, Pazzagli M, Bustin SA, Orlando C (2002) Quantitative real-time reverse transcription polymerase chain reaction: normalization to rRNA or single housekeeping genes is inappropriate for human tissue biopsies. Anal Biochem 309: 293–300
4. Selvey S, Thompson EW, Matthaei K, Lea RA, Irving MG, Griffiths LR (2001) Beta-actin–an unsuitable internal control for RT-PCR. Mol Cell Probes 15:307–311
5. Hamalainen HK, Tubman JC, Vikman S, Kyrola T, Ylikoski E, Warrington JA, Lahesmaa R (2001) Identification and validation of endogenous reference genes for expression profiling of T helper cell differentiation by quantitative real-time RT-PCR. Anal Biochem 299:63–70
6. Goidin D, Mamessier A, Staquet MJ, Schmitt D, Berthier-Vergnes O (2001) Ribosomal 18S RNA prevails over glyceraldehyde-3-phosphate dehydrogenase and beta-actin genes as internal standard for quantitative comparison of mRNA levels in invasive and noninvasive human melanoma cell subpopulations. Anal Biochem 295:17–21
7. Schmittgen TD, Zakrajsek BA (2000) Effect of experimental treatment on housekeeping gene expression: validation by real-time, quantitative RT-PCR. J Biochem Biophys Methods 46:69–81
8. Savonet V, Maenhaut C, Miot F, Pirson I (1997) Pitfalls in the use of several "housekeeping" genes as standards for quantitation of mRNA: the example of thyroid cells. Anal Biochem 247:165–167
9. Hurteau GJ, Spivack SD (2002) mRNA-specific reverse transcription-polymerase chain reaction from human tissue extracts. Anal Biochem 307:304–315
10. Eickhoff B, Korn B, Schick M, Poustka A, van der Bosch J (1999) Normalization of array hybridization experiments in differential gene expression analysis. Nucleic Acids Res 27:e33
11. Garbay B, Boue-Grabot E, Garret M (1996) Processed pseudogenes interfere with reverse transcriptase-polymerase chain reaction controls. Anal Biochem 237:157–159

12. Peeters RP, Wouters PJ, Kaptein E, van Toor H, Visser TJ, Van den Berghe G (2003) Reduced activation and increased inactivation of thyroid hormone in tissues of critically ill patients. J Clin Endocrinol Metab 88:3202–3211

13. Rutter J, Reick M, McKnight SL (2002) Metabolism and the control of circadian rhythms. Annu Rev Biochem 71:307–331

14. Jones LJ, Yue ST, Cheung CY, Singer VL (1998) RNA quantitation by fluorescence-based solution assay: RiboGreen reagent characterization. Anal Biochem 265:368–374.

15. Vincent VA, DeVoss JJ, Ryan HS, Murphy GM Jr. (2002) Analysis of neuronal gene expression with laser capture microdissection. J Neurosci Res 69:578–586

16. Solanas M, Moral R, Escrich E (2001) Unsuitability of using ribosomal RNA as loading control for Northern blot analyses related to the imbalance between messenger and ribosomal RNA content in rat mammary tumors. Anal Biochem 288 :99–102

17. Vandesompele J, De Preter K, Pattyn F, Poppe B, Van Roy N, De Paepe A, Speleman F (2002) Accurate normalization of real-time quantitative RT-PCR data by geometric averaging of multiple internal control genes. Genome Biol 3: RESEARCH0034

# The Choice of House Keeping Genes in MRD-Quantification of AML1-ETO Positive Acute Myeloid Leukemia

Martin Weisser*, Claudia Schoch, Torsten Haferlach,
Wolfgang Hiddemann, Susanne Schnittger

## Introduction

In modern hematology the quantification of minimal residual disease (MRD) during and after anti-leukemic therapy is of major interest. Reverse transcriptase polymerase chain reaction (RT-PCR) is the most sensitive and widely used method for detection and quantification of MRD. Common targets for MRD detection in acute myeloid leukemia (AML) are fusion transcripts, e.g., AML1-ETO, PML-RARα and CBFβ-MYH11, are specifically expressed by the leukemic clone (1). Several groups have detected molecular relapse of myeloid leukemias before hematologic manifestation by following ratios of fusion transcripts (2, 3, 4). Quantitative studies of RNA expression are usually performed relative to the expression of a housekeeping gene as an endogenous control that compensates for uneven RNA quality and different numbers of cells between samples. Housekeeping genes comprise a wide variety of genes that mainly code for proteins that are essential for cell function (5) and are assumed to be constitutively expressed in all cell types at relative stable levels. Recently, some reports have indicated that the expression of housekeeping genes can vary considerably depending on the experimental conditions (6, 7). Other studies have shown that neoplastic cells may exhibit abnormal levels of housekeeping gene expression (8, 9). Therefore, the choice of an appropriate housekeeping gene as an endogenous control is essential for the accuracy and reliability of a quantitative PCR approach. We have established a real-time RT-PCR protocol in our lab to quantify AML1-ETO fusion transcripts in t(8;21) positive AML patients. For this quantitative PCR approach, we examined the expression of four commonly used housekeeping genes – β-2-microglobulin (β2M), porphobilinogen deaminase (PBGD), glucose-6-phosphate dehydrogenase (G6PDH) and cABL – by real-time RT-PCR. Samples included bone marrow at primary diagnosis, follow-up, during and after cytostatic therapy, and the peripheral blood of healthy individuals.

* Dr. Martin Weisser, Department of Internal Medicine III, University of Munich, Klinikum Grosshadern, Marchioninistr. 15, 81377 Munich, Germany
E-mail: martin.weisser@med3.med.uni-muenchen.de

## Materials

LightCycler® instrument (Roche Diagnostics, Mannheim, Germany)
LightCycler® software, version 3.5 (Roche Diagnostics, Mannheim, Germany)
LightCycler® capillaries (Roche Diagnostics, Mannheim, Germany)
LightCycler® centrifuge (Roche Diagnostics, Mannheim, Germany)
EDTA stabilized human blood and bone marrow samples
Kasumi-1 cell line (10)

LightCycler FastStart DNA Master Hybridization Probes (Roche Diagnostics, Mannheim, Germany)
First Strand cDNA Synthesis Kit for RT-PCR (AMV) (Roche Diagnostics, Mannheim, Germany)
RNeasy® RNA-Isolation kit (Qiagen, Hilden, Germany)
Hybridization probes AML1-ETO, cABL, & G6PDH (TIB MOLBIOL, Berlin, Germany)
Primers AML1-ETO, cABL & G6PDH (TIB MOLBIOL, Berlin, Germany)
Housekeeping gene set β2-Microglobulin and PBGD (Roche Diagnostics, Mannheim, Germany)
Ficoll Hypaque density gradient (Sigma, Munich, Germany)

## Procedure

Human blood and bone marrow samples were analyzed. Peripheral blood of healthy individuals was taken from volunteers. Bone marrow samples of AML patients were referred to our lab for diagnosis and were selected due to t(8;21) positive cytogenetics. Mononuclear cells were obtained by standard Ficoll Hypaque density gradient centrifugation.

mRNA from $5 \times 10^6$ human cells was isolated according to the MagNA-Pure mRNA protocol for human cells (Roche Diagnostics, Mannheim, Germany).

cDNA Synthesis was performed via First Strand cDNA Synthesis Kit for RT-PCR (AMV) using random hexamers according to the manufacturer's protocol. A volume of 8 µl of mRNA corresponding to approximately $0.8 - 1.6 \times 10^6$ cells were transcribed to cDNA.

Real-time RT-PCR was performed using the LightCycler. For analysis of AML1-ETO, cABL and G6PDH, 2 µl of cDNA corresponding to $0.8 - 1.6 \times 10^5$ cells were amplified using the LightCycler Faststart DNA Master for Hybridization Probe reagents. The PCR products were detected via specific fluorescence-labeled hybridization probes. The detection format was based on fluorescence resonance energy transfer (FRET) (Wittwer 1997). The PCR conditions are shown in Table 1. PCR was performed using 2 µl mastermix containing buffer, dNTPs, and Taq polymerase, 4 mM $MgCl_2$; 0.25 µM of each 3´ and 5´ fluorescent-labeled hybridization probes; 0.5 µM of each 3´ and 5´ primer; 2 µl of cDNA, and $H_2O$ to a final volume of 20 µl. Analysis of β2M and PBGD was performed according to the manufacturer's protocol. LightCycler data was analyzed using LightCycler software, version 3.5. The fluorescence signal readings were analyzed using the second derivative maximum method.

Quantitative analysis of cABL, G6PDH, β2-M and PBGD expression in peripheral blood of healthy individuals was performed via standard curve analysis. Quantitative analysis of AML1-ETO fusion transcripts in the leukemia patients at primary diagnosis and during anti-leukemic therapy was normalized to these housekeeping genes. Relative expression levels of AML1-ETO fusion transcripts were calculated according to a known concentration of an external calibrator (mRNA of $10^5$ t(8;21) positive Kasumi-1 cell) as previously described (11). For every sample the expression level of AML1-ETO was normalized against the expression level of each housekeeping gene. These ratios were then correlated to each other.

**Table 1.** Reaction Mix (AML1-ETO, cABL, G6PDH)

|  | Volume [µl] | [Final] |
|---|---|---|
| LightCycler Faststart DNA Master | 2 | 1x |
| MgCl₂ (25 mM) | 2.4 | 4 mM |
| Probe mix (2.5 µM each) | 2 + 2 | 0.25 µM each |
| Primers (10 µM each) | 1 + 1 | 0.5 µM each |
| cDNA | 2 | |
| H₂O | 7.6 | |
| Total | 20 | |

**Table 2.** PCR conditions

| Parameter | Value | | |
|---|---|---|---|
| Cycles | 45 | | |
| Type | Quantification | | |
| | Segment 1 | Segment 2 | Segment 3 |
| Target temperature [°C] | 95 | 64 | 72 |
| Incubation time [s] | 10 | 10 | 26 |
| Temperature transition rate [°C/s] | 20 | 20 | 2 |
| Acquisition mode | None | Single | None |
| Fluorimeter gains | F1=1, F2=15 | | |

**Table 3.** Oligonucleotides

| AML1-ETO (GenBank Accession # D13979 | | | | | |
|---|---|---|---|---|---|
| | Position | Exon | Length | GC (%) | $T_m$ (°C) |
| **Primers** | | | | | |
| GAGGGAAAAGCTTCACTCTG | 2003 | 4 | 20 | 50 | 60.7 |
| TCGGGTGAAATGTCATTGCC | 2450R | 3 | 20 | 50 | 63.5 |
| **Probes** | | | | | |
| CCCTCACCACCCAATGGCTTCAGC-F | 2293 | 3 | 24 | 62.5 | 73.3 |
| LCRED640-TGGGCCTTCCTCTTCTTC-CTCCTCC-P | 2319 | 3 | 25 | 60 | 72.6 |
| **cABL (GenBank Accession # U07563)** | | | | | |
| **Primers** | | | | | |
| CCCAACCTTTTCGTTGCACTGT | 49996 | ABLa2 | 22 | 50 | 66.5 |
| CGGCTCTCGGAGGAGACGTAGA | 58610R | ABLa4 | 22 | 64 | 69.3 |
| **Probes** | | | | | |
| TGAAAAGCTCCGGGTCTTAGGC-TATAATCA-F | 50627 | ABLa3 | 30 | 43 | 70.1 |
| LCRED640-AATGGGGAATGGTGT-GAAGCCCAAA-P | 50658 | ABLa3 | 25 | 48 | 70.7 |
| **G6PDH (GenBank Accession # X55448)** | | | | | |
| **Primers** | | | | | |
| CCGGATCGACCACTACCTGGGCAAG | 15116 | 6 | 25 | 64 | 73.8 |
| GTTCCCCACGTACTGGCCCAGGACCA | 16443R | 9 | 26 | 65 | 76.8 |
| **Probes** | | | | | |
| GTTCCAGATGGGGCCGAAGATCCT-GTTG-F | 15376R | 7 | 28 | 57 | 72.7 |
| LCRED640-CAAATCTCAGCACCATGA-GGTTCTGCAC-P | 15165R | 7/6 | 28 | 50 | 70.0 |

**Table 4.** Gene expression (copies/1000 cells) of housekeeping genes in seven healthy volunteers

| | Mean | Range | Standard deviation | Coefficient of variation (%) |
|---|---|---|---|---|
| cABL | 20726 | 14879–27878 | 3375 | 16 |
| G6PDH | 14879 | 10220–19475 | 2801 | 19 |
| β2M | 2867200 | 2146000–3388000 | 444642 | 16 |
| PBGD | 2472 | 1355–4000 | 914 | 37 |

## Results

We quantified the gene expression levels of the four housekeeping genes in peripheral blood of seven healthy volunteers. cABL, G6PDH, β2M showed constant expression. In contrast, the expression level of PBGD was rather inconsistent in the peripheral blood of healthy volunteers (Figure 1-4). Thus, PBGD is probably not appropriate for MRD studies in AML1-ETO positive AML.

We analyzed the levels of AML1-ETO transcripts in seven t(8;21) positive AML patients at primary diagnosis. Relative quantification was performed by forming a ratio-of-target to housekeeping gene expression. The ratios of the corresponding samples were then correlated to each other. Quantification of AML1-ETO transcripts showed highest correlation when normalized against cABL, PBGD, and β2M as an endogenous control. Relative quantification using G6PDH showed a weaker correlation to cABL (R= 0.793) and β2M (R= 0.705) as shown in Table 5.

Quantitative Analysis of cABL, G6PDH, b2M and PBGD Expression in Seven Healthy Individuals

Analysis of Housekeeping Gene Expression in the Bone Marrow of Seven AML1-ETO Patients at Diagnosis

**Table 5.** Quantification of AML1-ETO fusion transcripts relative to the house keeping genes cABL, G6PDH, β2M and PBGD in seven patients at primary diagnosis

| | Coefficient of correlation ( r ) |
|---|---|
| cABL / G6PDH | 0.793 |
| cABL / β2M | 0.841 |
| cABL / PBGD | 0.929 |
| G6PDH / β2M | 0.705 |
| G6PDH / PBGD | 0.867 |
| β2M / PBGD | 0.846 |

We performed quantitative analysis of AML1-ETO fusion transcripts relative to the housekeeping genes cABL, G6PDH, β2M, and PBGD in seven follow-up bone marrow samples of two t(8;21) positive AML patients during the course of anti-leukemic therapy. Again a ratio-of-target to housekeeping gene was formed and the ratios of the corresponding samples were correlated. In the follow-up samples, during and after cytostatic therapy, relative quantification of the AML1-ETO fusion transcript revealed a highly significant correlation.

Relative Quantification of AML1-ETO Fusion Transcripts in Follow-Up Samples of t(8;21) Positive AML Patients

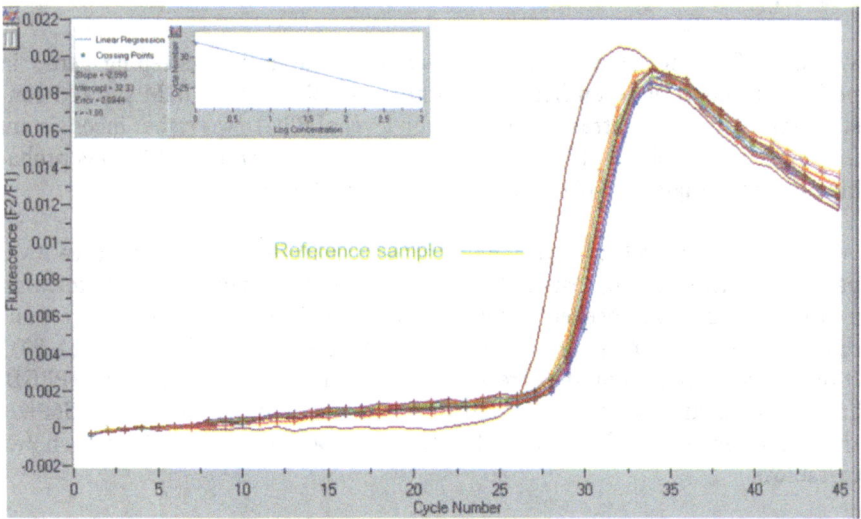

cABL

**Fig. 1**

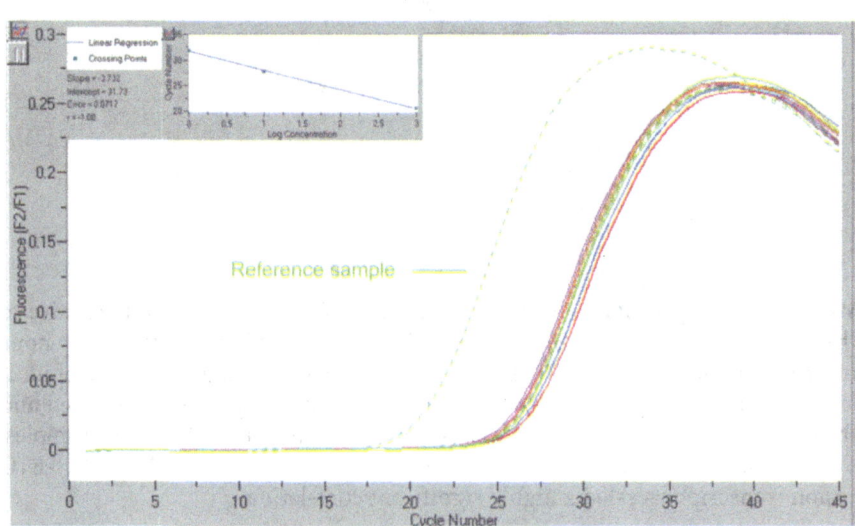

G6PDH

**Fig. 2**

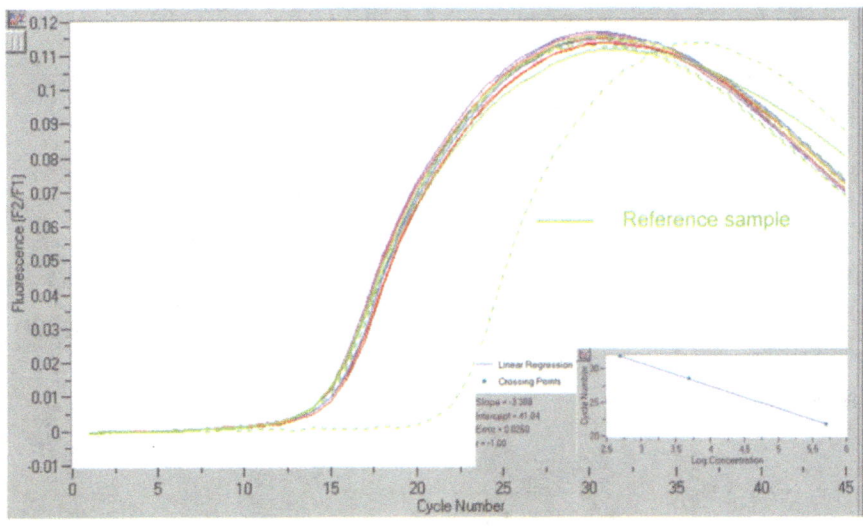

β2-M

**Fig. 3**

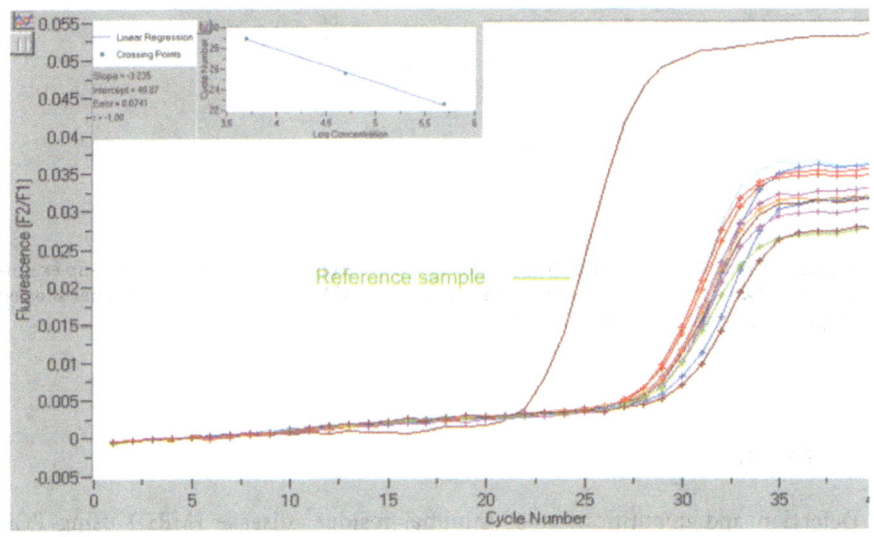

PBGD

**Fig. 4**

**Figures 1–4:** Amplification of cABL, G6PDH, β2M, PBGD in the peripheral blood of seven healthy individuals with corresponding standard curves and calibrator sample of a known concentration of cABL, G6PDH, β2M and PBGD transcripts

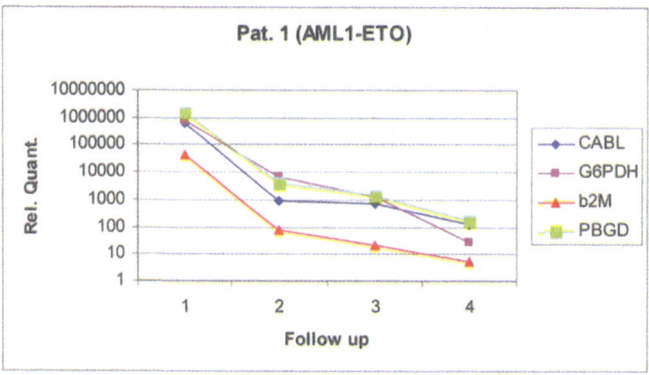

**Fig. 5**

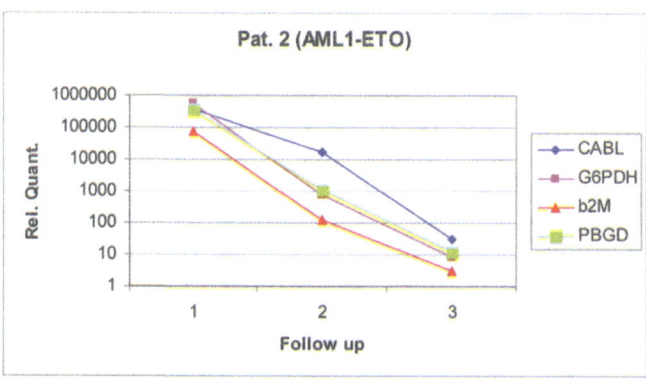

**Fig. 6**

**Figures 5 & 6:** Relative quantification of AML1-ETO transcript levels in seven follow-up examinations of two AML patients. Relative quantification of AML1-ETO to all control genes showed a highly significant correlation (R= 0.999)

## Discussion

Detection and quantification of minimal residual disease (MRD) using PCR based technologies is of increasing importance in the therapy of AML. For clinical application MRD detection needs to be informative, sensitive, reliable, and most important, quantitatively accurate. Furthermore, to compare results between laboratories, PCR protocols need to be standardized. A crucial point to establish a quantitative PCR application is the choice of an appropriate housekeeping gene as an endogenous control. Therefore, we compared the expression of four different housekeeping genes in AML1-ETO positive AML samples and healthy volunteers. We showed that cABL, G6PDH, and β2M are expressed at stable levels in the peripheral blood of healthy individuals. In contrast, the expression level of PBGD revealed higher variation than other tested control genes. Therefore, PBGD is not an ideal control gene for MRD studies as peripheral blood is often analyzed for post remission control. In t(8;21) positive AML patients quantification using cABL, β2M, and PBGD correlated well while the correlation with G6PDH showed the weakest correlation. In contrast, analysis of seven bone marrow samples of AML patients, during and after cytostatic therapy, revealed a highly significant correlation of expression level of all four control genes (r=0.99) indicating that all of the control genes can be used. However, we would not recommend PBGD, because of conflicting results of variance studies with peripheral blood of healthy individuals and the bone marrow of leukemia patients. G6PDH is not an ideal control gene for AML1-ETO quantification because of its weaker correlation compared to other tested housekeeping genes at primary diagnosis. In addition, β2M should be used with caution due to its very high expression level. High expression of the control gene may decrease the dynamic range of detection especially if the target gene is expressed at low levels or if the quality of the sample is low. In addition, there have been reports that β2M expression is altered in lymphoid malignancies which may hamper the use of this housekeeping gene in such entities (12, 13, 14). The remaining housekeeping gene, cABL, showed a level of expression that was in the range of the expression of the target AML-1-ETO at diagnosis, showed a constant homogenous expression in healthy volunteers, and high correlation in AML samples. In summary, we recommend using cABL for MRD studies in t(8;21) positive AML. Quantitative PCR protocols based on normalization against an endogenous control requires an appropriate housekeeping gene specific to the application.

# Reference

1. Liu Yin A, Grimwade D. (2002) Minimal residual disease evaluation in acute myeloid leukemia. Lancet; 360: 160–162
2. Schnittger S, Weisser M, Schoch C, Hiddemann W, Haferlach T, Kern W. (2003) New score predicting for prognosis in PML-RARA-, AML1-ETO-, or CBFB-MYH11- positive acute myeloid leukemia based on quantification of fusion transcripts. Blood. 2003 Oct 15;102(8):2746–55
3. Hochhaus A, Weisser A, La Rosee P, Emig M, Muller MC, Saussele S, Reiter A, Kuhn C, Berger U, Hehlmann R, Cross NC. Detection and quantification of residual disease in chronic myelogenous leukemia. Leukemia 2000, Jun;14(6):998–1005
4. Lo Coco F., Diverio D., Avvisati G. et al. (1999) Therapy of molecular relapse in acute promyelocytic leukemia. Blood 94:2225–2229
5. Singer, M. and Berg, P. (1991) Genes and Genomes. Oxford: blackwell Scientific Publications
6. Thellin O, Zorzi W, Lakaye B, De Borman B, Coumans B, Hennen G, Grisar T, Igout A, Heinen E. Housekeeping genes as internal standards: use and limits. J Biotechnol 1999 Oct 8;75(2–3):291–5
7. Schmittgen TD, Zakrajsek BA. Effect of experimental treatment on housekeeping gene expression: validation by real-time, quantitative RT-PCR. Biochem Biophys Methods 2000 Nov 20;46(1–2):69–81
8. Nicolson GL. Molecular mechanisms of cancer metastasis: tumor and host properties and the role of oncogenes and suppressor genes. Curr Opin Oncol 1991 Feb;3(1):75–92

9. Cox PM, Goding CR. Transcription and cancer. Br J Cancer 1991 May;63(5):651–62

10. Asou H, Tashiro S, Hamamoto K, Otsuji A, Kita K, Kamada N (1991) Establishment of a human acute myeloid leukemia cell line (Kasumi-1) with 8;21 chromosome translocation. Blood 77(9):2031–6

11. Weisser M., Schoch C., Haferlach T., Hiddemann W., Schnittger S. (2001) Quantitative Analysis of AML1-ETO Fusion Transcripts in t(8;21) positive AML Using Real-Time RT-PCR. Rapid cycle Real-Time PCR Methods and Applications, Genetics and Oncology, Springer-Verlag

12. Jones RA, Scott CS, Katz FE, Child JA. MHC class I and class I-like gene product expression by malignant T cells: relationships between CD1a, HLA-ABC and beta 2-microglobulin. Clin Exp Immunol 1988 Dec;74(3):454–8

13. Aviles A, Zepeda G, Diaz-Maqueo JC, Rodriguez L, Guzman R, Garcia EL, Talavera A. Beta 2 microglobulin level as an indicator of prognosis in diffuse large cell lymphoma. Leuk Lymphoma 1992 May;7(1–2):135–8

14. Coiffier B. Prognostic factors in Hodgkin's and non-Hodgkin's lymphomas. Curr Opin Oncol 1991 Oct;3(5):843–51

# Quantification of mRNA Using Linear Regression of Log-Linear PCR Data-Points as an Alternative for the Standard Curve Approach

Christian Ramakers[1], Daphne Timmer[2], Onno Bakker[2],
Ronald H. Lekanne Deprez[1], Jan M. Ruijter*[1], Antoon F.M. Moorman[1]

## Introduction

The real-time PCR technique is increasingly becoming the standard for quantification of messenger RNA (mRNA) levels in biomedical sciences [1,2]. This technique is preferred over other quantitative PCR methods because it allows high-throughput screening and, more importantly, does not rely on end-point analysis. The latter can be misleading due to product inhibition, increasing enzyme instability, and a depletion of reaction components with increasing cycle number [3].

A widely used analysis method for real-time PCR data relies on the construction of a standard curve. This method is based on determining the threshold cycle ($Ct$), which is the cycle number at which a fixed amount of DNA is formed for a series of samples with a known input. Standard curves are usually based on the $Ct$ values of either an input series of known RNA concentrations in the reverse transcription or a dilution series of a reference cDNA sample. The method uses the equation for PCR kinetics $N_{Ct}=N_0\times(Eff)^{Ct}$ in which $N_{Ct}$ is the fixed amount of product at threshold cycle $Ct$, $N_0$ is the starting concentration and $Eff$ is the efficiency of the PCR. This exponential equation is converted into a linear relation, $Ct=Log(N_{Ct})/Log(Eff)-1/Log(Eff)\times Log(N_0)$, between $Log(N_0)$ and $Ct$ in which $-1/Log(Eff)$ is the slope. Therefore a standard curve can be constructed by fitting a straight line to the $Log(N_0)$ and $Ct$ values for a known series of $N_0$. An important assumption in this standard curve method is that the PCR efficiency of the amplicon of interest is constant and has the same value in all standard curve and unknown samples. This assumption is, however, not always fulfilled.

In a recent publication we have shown that standard-curve-derived efficiencies for a GAPDH primer set can vary up to 20% due to random variations. To

* Jan M. Ruijter, Department of Anatomy & Embryology, K2–283, Academic Medical Center, University of Amsterdam, Meibergdreef 15, 1105 AZ, Amsterdam, The Netherlands
E-mail: j.m.ruijter@amc.uva.nl
[1] Experimental and Molecular Cardiology Group,
[2] Department of Experimental Endocrinology, Academic Medical Center, University of Amsterdam

circumvent $N_0$ quantification using a standard curve, we proposed an alternative method that uses linear regression in the Log-linear phase of individual samples [4]. In short, the exponential function $N_C=N_0×(Eff)^C$ is linearised by taking the logarithm on both sides of the equation: $Log(N_C)=Log(N_0)+Log(Eff)×C$. In this linear function, the slope term is represented by $Log(Eff)$ and the Y-axis intercept term by $Log(N_0)$. Note that this function is not valid in the plateau phase of the reaction. Using linear regression this equation can be fitted to the data points in the log-linear phase of each individual sample. The advantage of this approach is that it enables reliable quantification of $N_0$ values without using a standard curve, thereby avoiding the assumption of equal PCR efficiencies for all samples.

In this paper we demonstrate that when the PCR efficiencies of individual samples are more or less equal, both the linear regression and standard curve method can be used for quantifying $N_0$ values. However, when individual PCR efficiencies vary considerably, the standard curve method becomes unreliable, resulting in misinterpretation of starting concentrations, and in that case the linear regression method is the method of choice for accurate quantification.

## Materials

**Equipment**

LightCycler Instrument (Roche Diagnostics, Almere, The Netherlands)
LightCycler data analysis software version 3.532 (Roche Diagnostics)
LightCycler Capillary Tubes (Roche Diagnostics)
LinRegPCR software version 7.4 ([4]; bioinfo@amc.uva.nl, subject: LinRegPCR)

**Reagents**

High Pure RNA tissue Kit (Roche Diagnostics)
RiboGreen RNA Quantitation Kit (Molecular Probes, Leiden, The Netherlands)
First Strand cDNA synthesis kit with random primers (Roche Diagnostics)
DNA Master SYBR green I kit (Roche Diagnostics)

## Procedure

**Total RNA Isolation / First Strand Synthesis**

Total RNA was isolated from human liver samples using the High Pure RNA Tissue Kit (Roche Diagnostics) according to the manufacturer's protocol. With the RiboGreen RNA Quantitation Kit (Molecular Probes) the total RNA concentration in each sample was determined and subsequently all samples were diluted to 0.1 µg/µl. Single-stranded cDNA was obtained using 1 µg of total RNA and the First Strand cDNA synthesis kit with random primers (Roche Diagnostics).

**Table 1.** Human GAPDH [6] and EF-1α primers

| GAPDH (glyceraldehyde 3-phospate dehydrogenase) | | |
|---|---|---|
| Forward | TGA ACG GGA AGC TCA CTG G | |
| Reverse | TCC ACC ACC CTG TTG CTG TA | |
| $T_m$: 55 C | MgCl$_2$: 4 mM | Length amplicon: 307 base pairs |
| EF-1α (elongation factor-1α) | | |
| Forward | GAA CCA TCC AGG CCA AAT AA | |
| Reverse | CCG TTC TTC CAC CAC TGA TT | |
| $T_m$: 55 C | MgCl$_2$: 4 mM | Length amplicon: 392 base pairs |

**Table 2.** LightCycler reaction mix

| | Volume [µl] | [Final] |
|---|---|---|
| DNA Master SYBR green I | 2.0 | Containing 1 mM |
| H$_2$O (sterile, PCR grade) | 12.6 | |
| Forward primer (100 ng/µl) | 0.5 | 2.5 ng/µl |
| Reverse primer (100 ng/µl) | 0.5 | 2.5 ng/µl |
| MgCl$_2$ (25 mM) | 2.4 | 3 mM (total 4 mM) |

The mix was prepared for (n+1) samples. 18 µl was filled out in each capillary. Of each sample 2 µl cDNA was added to the individual capillaries

**Table 3.** LightCycler Quantification Report

| Program: Pre-incubation | | Type: quantification | | Cycles: 1 |
|---|---|---|---|---|
| Segment Number | Temperature target (°C) | Hold Time (s) | Slope (°C/s) | Acquisition Mode |
| 1 | 95 | 10 | 20 | None |
| **Program: Amplification** | | Type: quantification | | Cycles: 45 |
| Segment Number | Temperature target (°C) | Hold Time (s) | Slope (°C/s) | Acquisition Mode |
| 1 | 95 | 0 | 20 | None |
| 2 | 55 | 5 | 20 | None |
| 3 | 72 | 10 | 20 | Single |
| **Program: Melting** | | Type: melting curve | | Cycles: 1 |
| Segment Number | Temperature target (°C) | Hold Time (s) | Slope (°C/s) | Acquisition Mode |
| 1 | 95 | 0 | 20 | None |
| 2 | 70 | 15 | 20 | None |
| 3 | 95 | 0 | 0.1 | Continuous |
| **Program: Cooling** | | Type: none | | Cycles: 1 |
| Segment Number | Temperature target (°C) | Hold Time (s) | Slope (°C/s) | Acquisition Mode |
| 1 | 42 | 30 | 20 | None |

## Results

Specific amplification of the constitutively active genes EF-1α (elongation factor-1α), a protein involved in the translation of mRNA, and GAPDH (glyceraldehyde 3-phosphate dehydrogenase), an important enzyme in the glycolysis pathway, was confirmed by melting curve analysis revealing single melting peaks and poly-acrylamide gel electrophoresis showing single bands of the expected size for both amplicons (data not shown). The $N_0$ values were determined using the standard curve and linear regression methods for both amplicons and are given in table 4. Figure 1 shows the correlation between the $N_0$ values of both methods (linear regression vs. standard curve). We found a high degree of correlation for the $N_0$'s in the GAPDH data set (figure 1A, $R^2=0.92$), while the $N_0$'s in the EF-1α data set (figure 1B) did not show any correlation ($R^2=0.0025$).

In order to check the standard curves of GAPDH and EF-1α, the data were exported to Microsoft Excel™. The parameters of both standard curves are list-

**Table 4.** Individual PCR efficiencies for EF-1α and GAPDH in a sample population and a comparison of $N_0$ values from standard curve and linear regression methods. For standard curve quantification crossing points were determined with noise bands at 0.40 and 0.18 for EF-1α and GAPDH, respectively. Samples 1 and 13 (bold) are used as examples in the text and are shown in figure 4

| | EF-1α | | | | GAPDH | | | |
|---|---|---|---|---|---|---|---|---|
| | Standard curve | | Linear regression | | Standard curve | | Linear regression | |
| Sample nr. | Crossing point ($Ct$) | $N_0$ | $N_0$ ($\times10^{-4}$) | Ind. PCR Efficiency | Crossing point ($Ct$) | $N_0$ | $N_0$ ($\times10^{-5}$) | Ind. PCR Efficiency |
| 1 | **21.16** | **1.95** | **1.26** | **1.60** | **18.05** | **37.3** | **9.94** | **1.70** |
| 2 | 21.36 | 1.71 | 1.72 | 1.57 | 17.46 | 54.6 | 15.8 | 1.69 |
| 3 | 20.47 | 3.01 | 1.70 | 1.60 | 16.88 | 80.0 | 16.6 | 1.71 |
| 4 | 21.38 | 1.69 | 1.32 | 1.58 | 16.81 | 83.4 | 18.0 | 1.71 |
| 5 | 21.92 | 1.20 | 1.26 | 1.57 | 17.96 | 39.5 | 8.62 | 1.72 |
| 6 | 20.29 | 3.39 | 8.44 | 1.48 | 17.16 | 66.5 | 13.6 | 1.72 |
| 7 | 22.93 | 0.63 | 2.88 | 1.48 | 17.31 | 60.2 | 14.6 | 1.71 |
| 8 | 24.99 | 0.17 | 6.15 | 1.39 | 17.41 | 56.4 | 10.8 | 1.73 |
| 9 | 24.94 | 0.18 | 4.57 | 1.41 | 17.69 | 47.1 | 8.36 | 1.74 |
| 10 | 24.21 | 0.28 | 1.55 | 1.49 | 18.56 | 26.8 | 5.35 | 1.73 |
| 11 | 23.92 | 0.34 | 1.76 | 1.49 | 18.52 | 27.5 | 3.56 | 1.77 |
| 12 | 25.88 | 0.1 | 12.7 | 1.34 | 16.00 | 141 | 27.6 | 1.71 |
| 13 | **31.01** | **0.004** | **6.26** | **1.30** | **19.18** | **17.8** | **2.82** | **1.76** |
| 14 | 25.66 | 0.11 | 0.08 | 1.64 | 18.97 | 20.4 | 1.74 | 1.81 |
| 15 | 32.86 | 0.001 | 2.57 | 1.32 | 18.73 | 23.9 | 8.48 | 1.68 |
| 16 | 25.84 | 0.1 | 0.36 | 1.54 | 19.20 | 17.6 | 3.91 | 1.73 |
| 17 | 26.29 | 0.07 | 0.25 | 1.54 | 18.81 | 22.7 | 2.25 | 1.80 |
| 18 | 26.42 | 0.07 | 0.17 | 1.57 | 19.12 | 18.5 | 3.33 | 1.75 |

A

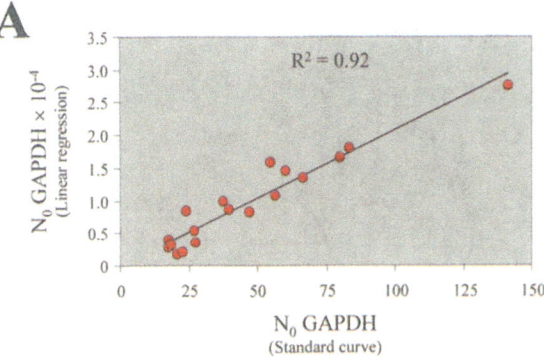

B

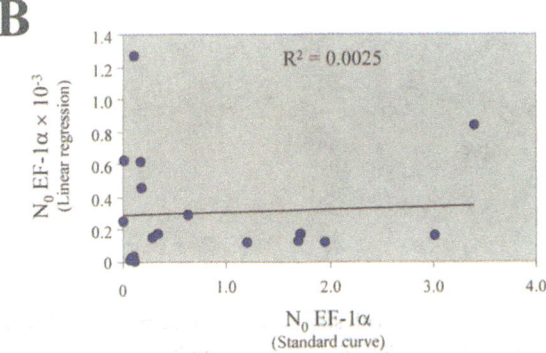

**Fig. 1.** Correlation between starting concentrations ($N_0$) obtained with linear regression (Y-axis; expressed as fluorescence units) and standard curve derived $N_0$ values (X-axis) for GAPDH (A) and EF-1α (B)

ed in figure 2A and B. The high correlation coefficient ($R^2$=1.00) of both curves showed a high degree of linearity between the $Ct$ and $Log(N_0)$ values. The standard-curve-derived efficiency was calculated as $10^{-1/slope}$. This resulted in efficiencies of 1.89 and 1.92 for EF-1α and GAPDH, respectively. The high degree of linearity and the observation that the standard-curve-derived PCR efficiencies had values lower than 2 (slope < –3.32) suggested that both standard curves were applicable for quantifying EF-1α and GAPDH levels in "unknown" samples [5].

Inspecting the individual PCR efficiencies of all samples (n=18) we found considerable variation in EF-1α efficiency (1.50 ± 0.10, mean ± sd) as shown in figure 3B and listed in table 4. The variation in the GAPDH data set was far less (figure 3A, table 4) with an average PCR efficiency of 1.73 ± 0.04. This indicates that the assumption of equal PCR efficiency of the standard curve method is not fulfilled for the EF-1α amplicon, which may explain the lack of correlation between the linear regression and standard curve $N_0$'s (figure 1A).

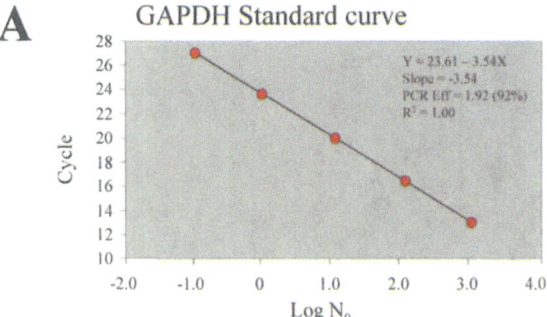

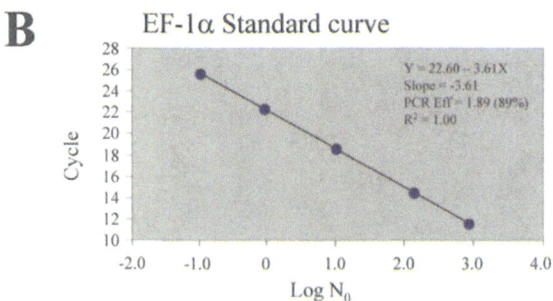

**Fig. 2.** GAPDH (A) and EF-1α (B) standard curves with standard curve parameters. Both standard curves are based on a serial dilution of the same purified cDNA sample

Within the EF-1α data set there were a couple of salient examples that show how quantification of samples with an aberrant PCR efficiency using a standard curve in combination with $Ct$ values can go wrong (figure 4). Figure 4A shows the EF-1α amplification curves of two samples (samples 1 and 13 from table 4). In the standard curve method, the crossing point of the noise band for sample 13 was reached after 31.01 cycles. Compared to most of the other $Ct$ values in table 4 this would indicate a very low expression level of EF-1α in this particular sample ($N_0$=0.004). Indeed, taking sample 1 as an arbitrary reference, the expression of EF-1α in sample 13 is ~500× **less** than in sample 1. However, this large fold-difference only occurs when the assumption is made that both PCR efficiencies are equal. Calculating the individual PCR efficiency of sample 13 showed a very low efficiency of 1.30 compared to 1.60 for sample 1. This low PCR efficiency of sample 13 explains why the crossing point of this sample was found at 31.01 cycles. In fact, when using the linear regression method for $N_0$ calculation, thereby taking the differences in individual PCR efficiency into account, the $N_0$ values showed ~5× **more** EF-1α transcript in sample 13 compared to sample 1. Comparing the same samples in the GAPDH data set (table 4, figure 4B) we found roughly the same number of fold-difference between samples 1 and 13 (standard curve: ~2× more; linear regression: ~3.5× more).

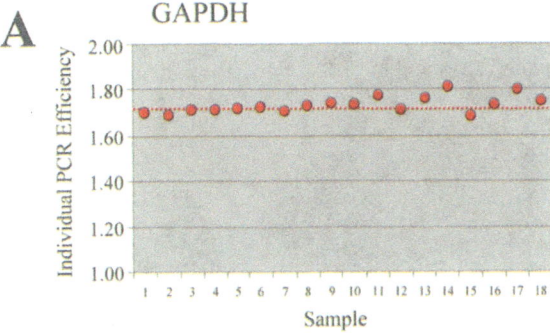

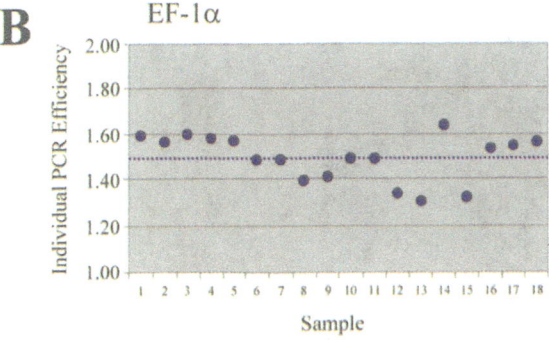

**Fig. 3.** Individual PCR efficiencies of GAPDH (A) and EF-1α (B) calculated using linear regression analysis of the log-linear data points of each amplification curve. The average PCR efficiencies (dotted lines) were 1.73 ± 0.04 (mean ± sd) and 1.50 ± 0.10 for GAPDH and EF-1α, respectively

## Comments

Based on the parameters of the standard curves of both EF-1α and GAPDH one would assume that both amplicons can be reliably measured. However, the $N_0$ values estimated with individual linear regression analysis of the EF-1α amplification curves do not correlate with those derived from the standard curve. This exemplifies the importance of taking the individual PCR efficiencies into account in quantitative PCR data analysis. When the PCR efficiencies are more or less equal (GAPDH, figure 3A), there is a high degree of correlation between the two $N_0$ values (figure 1A), suggesting that both methods can be reliably used for quantifying starting concentrations. In contrast, large variations in individual PCR efficiencies (EF-1α, figure 3B) result in a loss of correlation between the two $N_0$ values (figure 1B). We attribute this discrepancy to the inability of the standard curve method to deal with differences in individual sample PCR efficiencies, resulting in a misinterpretation of starting concentrations derived from this curve. Consequently, when the variation in individual PCR efficiencies is large, the

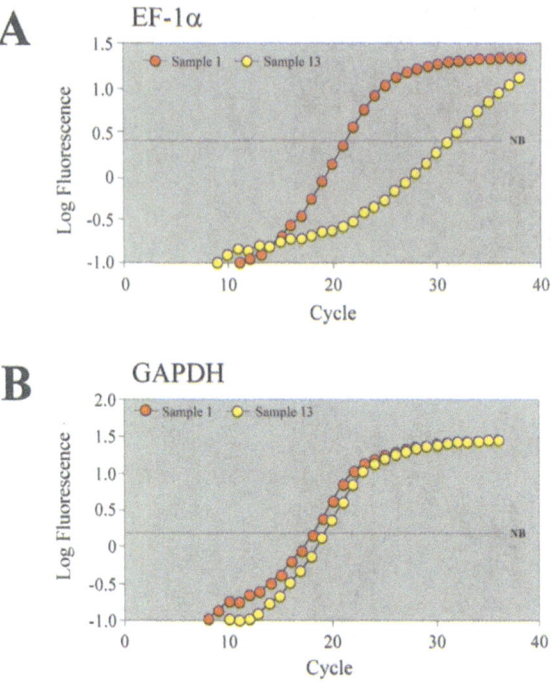

**Fig. 4.** Example showing the influence of individual PCR efficiencies on determining the $N_0$ values. The standard curve method uses a noise band (NB) to determine crossing points (*Ct* values) which in turn are used to calculate the $N_0$ values from the standard curve. For EF-1α (A) the noise band was set at 0.40. The noise band for GAPDH (B) was set at 0.18. The quantification parameters for the two samples are listed in table 4

$N_0$ values obtained with linear regression are more reliable than the standard curve $N_0$'s.

Different samples can have different PCR efficiencies. The cause of variation can be many, depending on for instance variation in reverse transcription efficiency, primer selection, (pharmacological) treatment of a patient prior to RNA isolation or leftover contaminants from tissue processing (e.g. ethanol, phenol). Unless a strict protocol for visual inspection of the log-linear amplification curves differences within and between LightCycler runs is employed in the laboratory, quantification of these samples using a standard curve can result in a gross misinterpretation of starting concentrations. Calculation of individual PCR efficiencies with the linear regression method allows the identification of samples with aberrant PCR efficiencies. Since all quantitative PCR data analyses assume that the PCR efficiency of a sample is constant for all cycles up to the end of the log-linear phase, the intercept of the regression line is a valid estimate of the $N_0$. Although this comparison of methods shows that $N_0$ calculations based on individual PCR efficiencies are more accurate when the variation in individual sample PCR efficiency is high, the question remains whether very low PCR efficiencies will give accurate quantification results. It may very well be that an efficiency of 30% for EF-1α sample 13 is too low for accurate quantification using either method. However, this example demonstrates rather dramatically how important individual PCR efficiencies are.

The fact that for the GAPDH experiment both methods give correlating results (figure 1A) indicates that the PCR efficiency variation for this amplicon does not violate the equal-PCR-efficiency assumption of the standard curve method. Nonetheless, we have previously shown that the slope of the standard curve is very dependent on small variations is PCR efficiencies of the samples used to construct the curve [4]. Whereas this would not greatly influence the relative differences of $N_0$ values of samples derived from one curve, its effect on $N_0$ results derived from different standard curves should not be neglected.

It is clear that differences in individual PCR efficiencies influence the $N_0$ values. With the standard curve method there is no possibility to correct for these differences. As mentioned before, linearization of $N_C=N_0 \times (Eff)^C$ not only yields an individual PCR efficiency, it also provides you with an Y-axis intercept, which is a direct measure of the $N_0$ value, thereby circumventing quantification with the standard curve assumption of equal PCR efficiencies. To facilitate this alternative quantification method we have developed a computer program that implements the linear regression analysis of real-time PCR using an automated algorithm [4]. This program is compatible with exported data from the LightCycler system (Roche), the ABI Prism systems (Applied Biosystems), the Opticon system (MJ Research) as well as the iCycler system (Biorad) and is available on request (*e-mail: bioinfo@amc.uva.nl; subject: LinRegPCR*).

# References

1. Gibson UEM, Heid CA, Williams PM (1996) A novel method for real-time quantitative RT-PCR. Genome Res 6: 995–1001
2. Wittwer CT, Ririe KM, Andrew RV, David DA, Gundry RA, Balis UJ (1997) The LightCycler™: A microvolume multisample fluorimeter with rapid temperature control. Biotechniques 22: 176–181
3. Kains P (2000) The PCR plateau phase – towards an understanding of its limitations. Biochim Biophys Acta 1494: 23–27
4. Ramakers C, Ruijter JM, Lekanne Deprez RH, Moorman AFM (2003) Assumption-free analysis of quantitative real-time PCR data. Neurosci Lett 339: 62–66
5. Lekanne Deprez RH, Fijnvandraat AC, Ruijter JM, Moorman AFM (2002) Sensitivity and accuracy of quantitative real-time PCR using SYBR Green I depend on cDNA synthesis conditions. Anal Biochem 307: 63–69
6. Kimura Y, Suzuki T, Kaneko C, Darnel AD, Moriya T, Suzuki S, Handa M, Ebina M, Nukiwa T, Sasano H (2002) Retinoid receptors in the developing human lung. Clin Sci 103: 613–621

# Measuring Genome Sizes by Absolute Quantification

Jochen Wilhelm, Meinhard Hahn*

## Introduction

The genome size (*i.e.*, the C-value) is defined as the amount of DNA in a haploid genome. Its exact determination is useful for phylogenetic studies, identifying species, and assessing the effort for sequencing projects on new species [1,2]. Determining the genomes size may seem easy, however, it is laborious and error prone in practice. Established techniques to determine genome sizes are based on measuring the phosphate content in the DNA from a defined number of cells, re-association kinetics of high molecular weight genomic DNA ($c_0t$-assays), flow cytometry, image analysis, or absorption cytometry after Feulgen staining. The method presented here is based on the absolute quantification of a single copy gene in a known amount (mass) of genomic DNA by real-time quantitative PCR. Absolute quantification is made possible by using a specific PCR product as an external standard whose concentration is determined by UV spectroscopy. The sample to be analyzed must contain pure DNA without significant RNA contamination. The C-value is calculated by dividing the mass of the DNA sample and by the copy number of the target gene, which is determined by absolute quantification on the LightCycler. This method was evaluated for *Homo sapiens sapiens* and proved to be very accurate. Results obtained with SYBR Green I and hybridization probes are compared.

### Materials

GeneAmp® PCR System 2400  (Applied Biosystems, Weiterstadt, Germany)  Equipment
LightCycler® instrument (Roche Diagnostics, Mannheim, Germany)
LightCycler® software, version 3.01 (Roche Diagnostics, Mannheim, Germany)
LightCycler® capillaries, centrifuge adapters, and cooling blocks (Roche Diagnostics, Mannheim, Germany)
OLIGO Primer Analysis software, version 5.0 (National Biosciences Inc., Plymouth, Minnesota, USA)
*So*FAR analysis software (own development) [3,4]
U-3000 UV-VIS spectrophotometer (Hitachi/Colora, Lorch, Germany)
SUPRASIL®-Quarzglas ultra-micro cuvette (Hellma, Müllheim/Baden, Germany)

---

* Meinhard Hahn, Abteilung Molekulare Genetik, B060, Deutsches Krebsforschungszentrum, Im Neuenheimer Feld 280, D-69120 Heidelberg, Germany. E-mail: m.hahn@dkfz.de

**Samples**
Human genomic DNA was isolated from lymphocytes of fresh, EDTA-treated blood, taken from healthy Caucasian individuals.

**Reagents**
QIAamp® DNA Mini Kit (QIAGEN, Hilden, Germany)
QIAquick® PCR Purification Kit (QIAGEN, Hilden, Germany)
Deoxynucleotides (dATP, dCTP, dGTP, dTTP) (Roche Diagnostics, Mannheim, Germany)
*Taq* DNA polymerase (Roche Diagnostics, Mannheim, Germany)
10x reaction buffer (100 mM Tris-HCl, 15 mM $MgCl_2$, 500 mM KCl, pH 8.3 at 20°C) (Roche Diagnostics, Mannheim, Germany)
Bovine serum albumin (BSA; molecular biology grade) (Roche Diagnostics, Mannheim, Germany)
$MgCl_2$, 25 mM (Roche Diagnostics, Mannheim, Germany)
SYBR Green I solution (10 000x in DMSO) (Roche Diagnostics, Mannheim, Germany)
PCR primers, HPSF grade (MWG-Biotech, Ebersberg, Germany)
Hybridization probes (TIB MOLBIOL, Berlin, Germany).

## Procedure

An overview of the experimental setup is shown in Figure 1. The procedure can be outlined as follows: a genomic DNA sample with a known DNA concentration is used to generate a specific PCR product for a single copy gene. This PCR product is purified and quantified by UV absorption spectroscopy. Subsequently, this product is used as an external standard in quantitative real-time PCR where the target gene copy number in the genomic DNA sample is estimated.

**Sample Preparation**
DNA from human blood was isolated using the QIAamp DNA Blood Mini Kit following the manufacturer's instructions, also including an extensive incubation step with RNase A.

Quality and concentration of all DNA samples were determined by recording UV absorption spectra between 220 and 320 nm with a spectrophotometer. The spectra were corrected for the absorption at 320 nm. A DNA solution with one absorption unit at 260 nm ($OD_{260}$) had a concentration of 50 ng/µl double-strand DNA (dsDNA). The DNA purity was acceptable when the ratio $OD_{260}$:$OD_{280}$ was between 1.8 and 1.9. In addition, the shape of the spectra was used to confirm the absence of light scattering particles and organic compounds that might impair the absorbance at 260 nm. The concentration of the purified DNA solution was adjusted to 15.0 ng/µl.

**Oligonucleotides**
Primer and probe (see Table 1) design were based on GenBank entry X54156 for the human gene *p53* using the OLIGO Primer Analysis software. The amplified region comprises the sequence segment between positions 14,513–14,854 (342 bp) for the outer primer pair, and 14,674–14,798 (125 bp) for the inner primer pair.

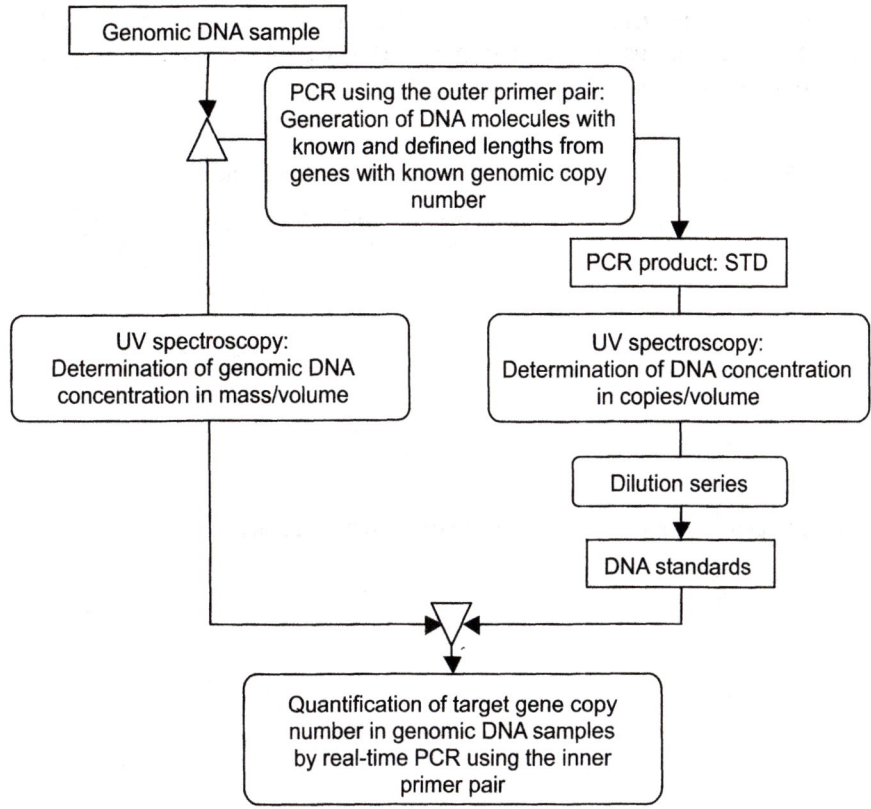

**Fig. 1.** Schematic of the experimental setup

**Table 1.** Oligonucleotides

| p53 gene of *H. sapiens sapiens* | | | | |
|---|---|---|---|---|
| Sequence 5'-3' | Length | GC(%) | $T_m$ (°C) | Purity |
| **Outer primers** | | | | |
| CGGCGCACAGAGGAAGAGAAT | 21 | 57.1 | 66.18 | 1.72 |
| CAAATGCCCCAATTGCAGGTA | 21 | 47.6 | 62.93 | 1.69 |
| **Inner primers** | | | | |
| TTCCTAGCACTGCCCAACA | 19 | 52.6 | 62.76 | 1.83 |
| GACTGGAAACTTTCCACTTG | 20 | 45.0 | 57.99 | 1.88 |
| **Probes** | | | | |
| CCCCAGCCAAAGAAGAAACCACTGG-ATGGAGAAT-F | 34 | 48.6 | 72.47 | 1.02 |
| LCRed640-TTTCACCCTTCAGGTACTA-AGTCTTGGGACCTCTT-p | 35 | 45.7 | 71.17 | 0.81 |

Generation of
Standards The outer primers were used to generate the standards in a conventional PCR from genomic DNA, using a GeneAmp PCR 2400 system and the following reaction conditions.

Table 2. PCR mix to generate the standard PCR products in 25 µl reactions

|  | Volume [µl] | [Final] |
| --- | --- | --- |
| Primers (5 µM each) | 2.5 | 0.5 µM each |
| dNTPs (2 mM each) | 2.5 | 200 µM each |
| Taq DNA polymerase (5 U/µl) | 0.5 | 2.5 U |
| Reaction buffer (10 x) | 2.5 | 1 x |
| Template (15 ng/µl genomic DNA) | 4.0 | 60 ng |
| Water (PCR grade) | 13.0 | |

Table 3. Amplification protocol to generate the standard PCR products

| Parameter | Value | | |
| --- | --- | --- | --- |
| Cycles | 35 | | |
|  | Segment 1 | Segment 2 | Segment 3 |
| Target temperature [°C] | 95 | 55 | 72 |
| Incubation time [s] | 30 | 30 | 60 |

The outer PCR product was purified using the QIAquick PCR Purification kit. The PCR product was eluted with water. Concentration and quality of the PCR product was determined by an UV absorbance spectrum. The absence of unspecific products was confirmed by electrophoretic separation in nondenaturing 12% polyacrylamide gels and final ethidium bromide staining.

Knowing the product length (342 bp) and the $OD_{260}$, the concentration (copies per volume) can be calculated by

$$c = \frac{OD_{260} \times 5 \times 10^{-8} \, \frac{g}{\mu l}}{[bp] \times \frac{660 \, \frac{g}{mol \cdot bp}}{6.022 \times 10^{23} \, mol^{-1}}}$$

The average molecular weight of 1 bp was assumed to be 660 g/mol. The concentration of the purified PCR products was adjusted to $5 \times 10^7$ copies per microliter.

LightCycler PCR The quantitative LightCycler PCR was performed using the inner primer pair. The purified genomic DNA target (15 ng/µl) served as sample. The human single copy gene *p53* was quantified with external standardization, using the previously generated, purified and quantified gene specific outer PCR product as standard DNA (1:10 dilution series with $5\times10^7$ to $5\times10^2$ copies per microliter). All pipeting steps were performed at 4°C, the LightCycler capillaries were pre-chilled in a cooled block with centrifuge adapters.

**Table 4.** LightCycler PCR master mix for each 10-μl reaction to quantify the copy number of a single copy gene using SYBR Green I

|  | Volume [μl] | [Final] |
|---|---|---|
| Primers (5 μM each) | 1.00 | 0.5 μM each |
| dNTPs (2 mM each) | 1.00 | 200 μM each |
| *Taq* DNA polymerase (5 U/μl) | 0.10 | 0.5 U |
| BSA (20 g/l) | 0.25 | 0.5 g/l |
| MgCl$_2$ (25 mM) | 1.80 | 6 mM |
| SYBR Green I (1:300) | 0.10 | 1: 30,000 |
| Reaction buffer (10 x) | 1.00 | 1 x |
| Water (PCR grade) | 4.75 | |

**Table 5.** Amplification program for SYBR Green I detection with the LightCycler

| Parameter | Value | | |
|---|---|---|---|
| Cycles | 45 | | |
| Analysis mode | Quantification | | |
| | Segment 1 | Segment 2 | Segment 3 |
| Target temperature [°C] | 95 | 55 | 72 |
| Incubation time [s] | 0 | 5 | 10 |
| Temperature transition rate [°C/s] | 20 | 20 | 20 |
| Acquisition mode | None | None | Single |
| Gains | F1 = 5 | | |

**Table 6.** Melting curves program for SYBR Green I detection with the LightCycler

| Parameter | Value | | |
|---|---|---|---|
| Cycles | 1 | | |
| Analysis mode | Melting curve | | |
| | Segment 1 | Segment 2 | Segment 3 |
| Target temperature [°C] | 72 | 95 | 40 |
| Incubation time [s] | 10 | 0 | 10 |
| Temperature transition rate [°C/s] | 20 | 0.1 | 20 |
| Acquisition mode | None | Continuous | None |
| Gains | F1 = 5; F2 = 15 | | |

**Table 7.** LightCycler PCR master mix for each 10-µl reaction to quantify the copy number of a single copy gene using hybridization probes

| | Volume [µl] | [Final] |
|---|---|---|
| Primers (5 µM each) | 1.00 | 0.5 µM each |
| dNTPs (2 mM each) | 1.00 | 200 µM each |
| *Taq* DNA polymerase (5 U/µl) | 0.10 | 0.5 U |
| BSA (20 g/l) | 0.25 | 0.5 g/l |
| MgCl$_2$ (25 mM) | 1.80 | 6 mM |
| Fluorescein-labeled probe (2 µm) | 1.00 | 0.2 µM |
| LCRed640-labeled probe (4 µM) | 1.00 | 0.4 µM |
| Reaction buffer (10 x) | 1.00 | 1 x |
| Water (PCR grade) | 2.85 | |

**Table 8.** Amplification program for hybridization probes detection with the LightCycler

| Parameter | Value | | |
|---|---|---|---|
| Cycles | 45 | | |
| Analysis mode | Quantification | | |
| | Segment 1 | Segment 2 | Segment 3 |
| Target temperature [°C] | 95 | 55 | 72 |
| Incubation time [s] | 0 | 5 | 10 |
| Temperature transition rate [°C/s] | 20 | 20 | 20 |
| Acquisition mode | None | Single | None |
| Gains | F1 = 5; F2 = 15 | | |

The LightCycler data were evaluated with the software *So*FAR [3], using the preset adjustments (background correction, sigmoid amplification function, and automatic threshold determination). The data recorded in channel 1 were used for SYBR Green I experiments, the ratio of the data from channel 2 and channel 1 were used for the hybridization probes experiments.

**Calculations**   The DNA amount that corresponds to the haploid genome (C-value) is derived from the ratio of the DNA template mass and the target sequence copy number:

$$C = \frac{m}{n}$$

m: mass of the DNA template as determined by UV absorbance at 260 nm;
n: number of target gene copies as determined by LightCycler PCR.

The C-value can be expressed in base pairs per haploid genome, also named $\Gamma$, by

$$\Gamma = C \cdot \frac{N_A}{M_{bp}}$$

$N_A$: Avogadro's number ($6.022 \cdot 10^{23}$ mol$^{-1}$)
$M_{bp}$: mean molar mass of a base pair (660 g·mol$^{-1}$)

## Results

Figure 2 shows the gel electrophoretic analysis results of PCR products obtained for the different primer/template combinations after 45 cycles of amplification. In each case, only the specific product was amplified. The non-template control (NTC) shows a minor accumulation of primer dimers.

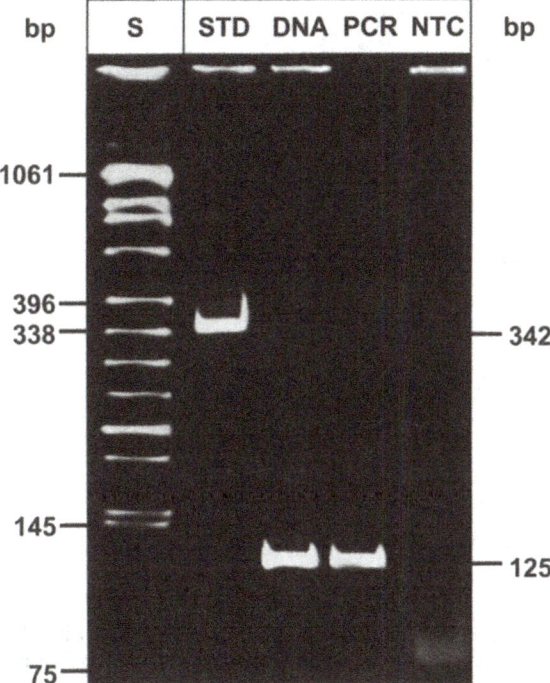

**Fig. 2.** Gel electrophoresis of the PCR products. Two microliters of each PCR reaction were analyzed on a non-denaturing 15% polyacrylamide gel. The gel was stained with ethidium bromide and an image was taken by a CCD camera after illumination with light of 254 nm.
S: size standard (pTR54 x *Hinf*III), **STD**: PCR product of the outer primers (used as standard in real-time PCR), **DNA**: real-time PCR product using genomic DNA as template, **PCR**: real-time PCR product using STD template, **NTC**: no template control with inner primer pair

The amplification curves for SYBR Green I and for the hybridization probes are shown in Figure 3.

Both detection formats gave very similar results. For SYBR Green I detection, the NTC sample shows the beginning of an exponential increase after 40 cycles due to the accumulation of primer dimers (see lane "NTC" in Figure 2). This was confirmed for real-time PCR products by a melting curve analysis (Figure 4).

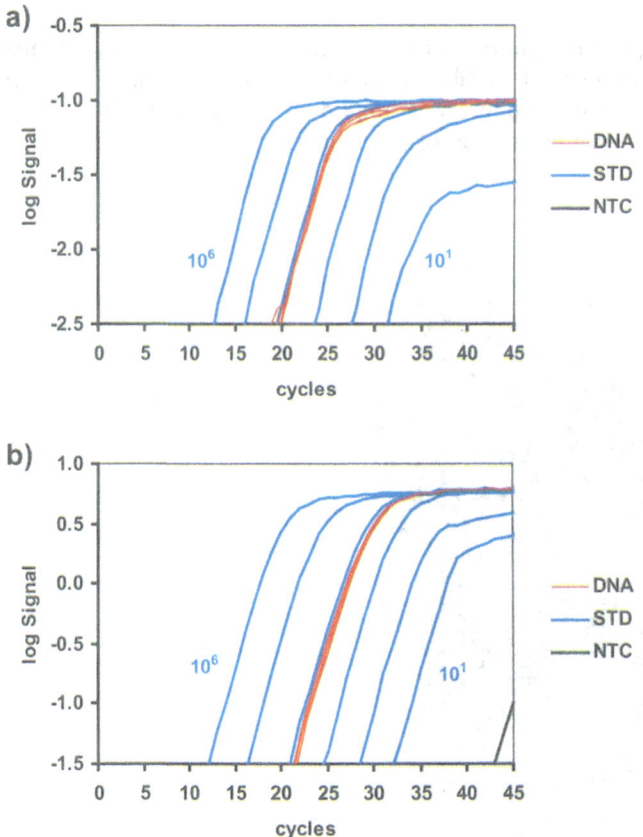

**Fig. 3.** Amplification curves. **a)** Detection with hybridization probes. **b)** Detection with SYBR Green I. Note the different y-axis intercepts in a) and b). In both diagrams, the amplification curves have comparable shapes, but for SYBR Green I, the maximum fluorescence yield is more than 50 times higher than for hybridization probes. Due to the logarithmic scale of the y-axis, the exponential phases of the amplification curves appear as straight segments. These segments are equidistant for the standard curve dilutions (STD). The curves of samples amplified from genomic DNA (DNA) have straight segments parallel to those of the standards, indicating that they are amplified with the identical efficiency.

**DNA:** sample with 30 ng genomic DNA; **STD:** standards containing $10^6$ to $10^1$ copies (1:10 dilution series) of the amplicon generated with OPP; **NTC:** no template control

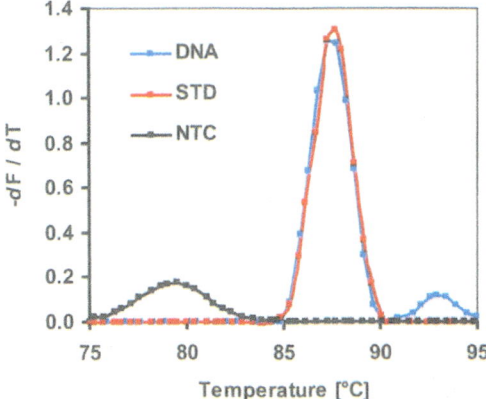

**Fig. 4.** Melting curves. Shown are the temperature quench corrected negative derivatives of the fluorescence in channel 1 (F) by the temperature (T). The faster the fluorescence decreases with increasing temperature, the higher is the value of the melting curve. The $T_m$ of a product is defined as the temperature of the steepest decrease of fluorescence, *i.e.*, the peaks of the melting curves. The temperature quench correction removes the effects of temperature dependent loss of fluorescence intensity. The areas under the curves are proportional to the amount of product. **DNA:** genomic DNA as template (sample); **STD:** amplicon generated with OPP used as template (standard); **NTC:** no template control

The specific product for the inner primer pair melts at 87.5°C. The $T_m$ is slightly lower (87°C) when genomic DNA is used as a template. This difference may be due to reduced free $Mg^{2+}$ concentration, because a significant amount of magnesium ions is bound to the genomic DNA. The melting peak at 93°C is regularly seen when genomic DNA is used as a template and may be due to the melting of the high molecular weight genomic DNA. The primer dimers' melting temperature is approximately 79.5°C.

Detailed results for all experiments are summarized in Table 9. For both detection formats, the estimated size of the haploid human genome is 3.2 pg or 2.92 Gbp. This value is consistent with the values derived from new textbooks, which propose a size between 2.8 and 3.2 Gbp (*e.g.,* [5]). By sequencing the human genome, Venter and co-workers calculated a size of 2.916 Gbp for the human genome [6], which exactly matched the value determined here by quantitative real-time PCR.

## Comments

The closed-tube format of real-time PCR analysis considerably reduces the risk of carry-over contamination of PCR products. In the protocol described here, PCR products must be generated in a standard PCR reaction. These purified products are subsequently used as standard DNA for an external calibration. This repre-

**Table 9.** Quantification experiment results

| | Hybridization probes | SYBR Green I |
|---|---|---|
| **Calibration line** | | |
| Slope (m) | −3.78 | −3.88 |
| Intercept (b) | 35.88 | 38.08 |
| Coefficient of correlation ($R^2$) | 0.9998 | 0.9999 |
| Efficiency | 1.84 | 1.81 |
| **Results** (copies per microliter) | | |
| #1 | 9282 | 9331 |
| #2 | 9551 | 9322 |
| #3 | 9270 | 9483 |
| #4 | 9418 | 9561 |
| #5 | 9367 | 9171 |
| Mean (copies per microliter) | 9378 | 9374 |
| Coefficient of variation (%) | 1.22 | 1.62 |
| C-value (pg) | 3.2 | 3.2 |
| Γ-value (Gbp) | 2.92 | 2.92 |

sents a very considerable source of contamination. Thus, the (diluted) standards should be prepared in a different lab.

There is no significant difference in the results obtained using either hybridization probes or SYBR Green I for detection. The accumulation of primer dimers can only be recognized from the NTC amplification when SYBR Green I is used. In addition, SYBR Green I provides a very convenient way to estimate product specificity. The quantification results are only reliable when no unspecific products have occurred during amplification. Otherwise, the amplification efficiency for the particular reaction might be reduced, leading to an increased $C_T$ value and an overestimate of the genome size. Because SYBR Green I is considerably less expensive than hybridization probes and provides more information about the occurrence of unspecific products, we prefer the SYBR Green I detection format over hybridization probes for quantification with external calibration. In addition, this technique has been validated by analysis of the genome size of *Saccharomyces cerevisiae* and *Xiphophorus maculatus* [7].

# References

1. Tiersch TR, Chandler RW, Wachtel SS, Elias S (1989) Reference standards for flow cytometry and application in comparative studies of nuclear DNA content. Cytometry 10:706–710
2. Shapiro HS (1970) Nucleic Acids. In: CRC Handbook of Biochemistry. Selected Data for Molecular Biology (Sober HA, ed): H-113. CRC Press, Cleveland, Ohio, USA
3. Wilhelm J, Pingoud A, Hahn M (2003) SoFAR: Software for fully automatic and highly accurate evaluation of real-time PCR data. BioTechniques 34:324–332
4. Wilhelm J, Pingoud A, Hahn M (2003) Validation of an algorithm for automatic quantification by real-time PCR. Anal Biochem 317:218–225

5.  Knippers R (2001) Molekulare Genetik. 8th edn. Georg Thieme Verlag, Stuttgart
6.  Venter JC, Adams MD, Myers EW, Li PW, Mural RJ, Sutton GG, Smith HO, Yandell M, Evans CA, Holt RA *et al.*(2001) The sequence of the human genome. Science 291:1304–1351
7.  Wilhelm J, Pingoud A, Hahn M (2003) Real-time PCR-based method for the estimation of genome sizes. Nucleic Acids Res 31:e56

# Applications

## Regulation and Development

II

# Relative Quantification of Insulin Gene Expression on the LightCycler Using SYBR Green I

NICOLE NEUBAUER*

## Introduction

Diabetes is characterized by dysfunction of the insulin-producing cells. The intestinal hormone glucagon-like petide-1 (GLP-1) is known to stimulate pancreatic beta cells in terms of insulin secretion, insulin gene expression and cell proliferation. It has been suggested, that defects in the GLP-1 signaling pathways within the beta cells are involved in the development of type 2 diabetes. To determine the stimulatory effects of GLP-1 on the insulin gene expression *in vitro* we decided to use real-time PCR rather then Northern blot [1], as we have to deal with limited amount of starting material when working with islets of Langerhans isolated from rat pancreas. Furthermore, the high sensitivity of this method is crucial to determine physiologically relevant changes in insulin gene expression. Expression levels are quantified relative to control samples and normalized to the expression of a housekeeping gene.

For our approach we used SYBR Green I for detection of dsDNA amplified in the PCR. SYBR green I dye binds to the minor groove of double stranded DNA. When bound to DNA the fluorescence of the dye is greatly enhanced and thereby the increase of the fluorescence during the PCR corresponds to the increased amounts of dsDNA that is amplified. Crucial for an application using SYBR green I is the amplification of the sequence of interest without any unspecific products. This objective requires a good primer design.

Conventional end-point PCR assays can be directly transferred to the real-time PCR system. As with the hybridization probes, the quantification of starting material is possible using SYBR green I. The advantage over the end-point PCR is the accuracy of the real-time technology because the exponential phase of the reaction will be visualized for each sample.

## Materials

LightCycler Instrument (Roche Diagnostics, Mannheim, Germany)
Mastercycler gradient (Eppendorf, Hamburg, Germany)
BioPhotometer (Eppendorf, Hamburg, Germany)

---

* Nicole Neubauer, Dipl. Biol., Ph.D. student, Department of Medical Biochemistry and Genetics, Panum Institute, University of Copenhagen, Blegdamsvej 3, 2200 Copenhagen N Denmark
E-mail: nicole@imbg.ku.dk

QIAshredder (Qiagen, Hilden, Germany)
RNeasy Mini Kit (Qiagen, Hilden, Germany)
RQ1 RNase-Free DNase (Promega, Madison, USA)
Reverse Transcription System (Promega, Madison, USA)
LightCycler FastStart Dann Master SYBR Green I (Roche)

Human serum (Novo Nordisk A/S, Gentofte, Denmark)
Human Glucagon like peptide-1 (Bachem AG, Weil am Rhein, Germany)
Human Growth hormone (Novo Nordisk A/S, Gentofte, Denmark)
Glutamax (Invitrogen A/S, Taastrup, Denmark)

## Procedure

For all optimization reactions as well as for efficiency determination of the PCR, total RNA from the insulinoma cell line INS-1E (kindly provided by Dr. C. Wollheim, University of Geneva) was used. The cell line was cultured as described elsewhere [2] and preparation of cDNA was performed as described with islet material.

Pancreatic islets were isolated from newborn Wistar rats by the collagenase method [3]. The islets were precultured at 37°C in RPMI 1640 containing 11 mM glucose, 100 units/ml penicillin, 100 µg/ml streptomycin, 7.5 % $NaHCO_3$, 1% glutamax, 2% human serum and 500 ng/ml human growth hormone. The islets were incubated for 7 days to allow attachment [4]. Prior to stimulation, the medium was changed and the islets were incubated in RPMI 1640 containing 3.3 mM glucose, 0.5 % human serum without human growth hormone over night. The islets were stimulated with 10 nM GLP-1 for 3 h, harvested and total RNA was extracted using the RNeasy Mini Kit. Total RNA content of the eluate was determined measuring optical density at 260 nm and purity by a 260/280 nm absorption ratio > 1.8.

To avoid any DNA contamination samples were treated with RNase free DNase (Promega) as described in the protocol of the manufacturer. Equal amounts of total RNA (0.5 µg) were treated and the final reaction volume was 11 µl for each sample.

The reverse transcription was performed following the manufactures protocol at 42°C for 15 minutes. DNase treated sample material (9.75 µl) were reverse transcribed in a total reaction volume of 20 µl containing: 5 mM $MgCl_2$, 10 mM Tris-HCl, 50 mM KCl, 0.1 % Triton ® X-100, 1 mM each dNTP, 1 unit/µl Recombinant RNasin® Ribonuclease Inhibitor, 15 units AMV Reverse transcriptase and 0.5 µg random primers.

Primers designed for analyzing the insulin gene expression amplify a fragment of 139 bp spanning two exons. The reverse primer is spanning an intron. Primers were designed to anneal at about 58°C to combine them with tested primers for the reference gene TATA box binding protein (TBP). The TBP primers amplify a fragment of 190 bps (Table 1).

**Table 1.** Oligonucleotides

| TATA Box binding protein (GenBank Accession # BC012685) | | | | |
|---|---|---|---|---|
| | Position | Length | GC (%) | $T_m$ (°C) |
| ACCCTTCACCAATGACTCCTATG | 830 | 23 | 47.8 | 63.0 |
| ATGATGACTGCAGCAAATCGC | 1019 | 21 | 47.6 | 63.3 |
| Amplification fragment from rat mRNA | 830–1019 | 190 | | |
| Insulin (GenBank Accession # NM_019130) | | | | |
| TGCCCAGGCTTTTGTCA | 63 | 17 | 52.9 | 60.8 |
| CTCCAGTTGTGCCACTTGT | 201 | 19 | 52.6 | 61.7 |
| Amplification fragment from rat mRNA | 63–201 | 139 | | |

LightCycler reaction mix:

Each 20 µl reaction contained 2 µl template cDNA, 0.5 µM of each gene-specific primer and previously determined optimal $MgCl_2$ concentration for each reaction. FastStart DNA Master SYBR Green I mix was used containing the hot start modified Taq DNA polymerase.

The real-time PCR program included a 10 min denaturation step to activate the Taq DNA Polymerase followed by a three-step amplification program described in Table 2.

After each LightCycler run a melting curve analysis was performed to analyse the products of the PCR (Table 3).

LightCycler PCR products were separated by agarose gel electrophoresis once to confirm the right amplification products. In later runs the melting curve analysis was sufficient to verify the proper fragment amplification.

A standard curve was performed for insulin and TBP by dilution of cDNA to determine the efficiency of the target and the reference reaction. The cDNA for these dilution curves was generated from the insulinoma cell line INS-1E. The cDNA from INS-1E was prepared identical to the cDNA from rat islets.

Both standard curve files were exported and used in the RelQuant program from Roche to create a coefficient file.

**Table 2.** Amplification program for 40 cycles:

| Parameter | Value | | |
|---|---|---|---|
| Cycles | 40 | | |
| Type | Amplification | | |
| Temp. Targets | Segment 1 | Segment 2 | Segment 3 |
| Target temperature [°C] | 95 | 58 | 72 |
| Incubation time [s] | 10 | 5 | 20 |
| Temperature transition rate [°C/s] | 20 | 20 | 20 |
| Acquisition mode | None | None | Single |

**Table 3.** Melting Curve Analysis (single cycle):

| Parameter | Value | | |
|---|---|---|---|
| Cycles | 1 | | |
| Type | Melting Curve | | |
| Temp. Targets | Segment 1 | Segment 2 | Segment 3 |
| Target temperature [°C] | 95 | 65 | 95 |
| Incubation time [s] | 0 | 15 | 0 |
| Temperature transition rate [°C/s] | 20 | 20 | 0.1 |
| Acquisition mode | None | None | Cont. |

For quantification of insulin gene transcription in GLP-1 stimulated rat islet, triplicates of control and stimulated samples were run and the data exported as standard curve data from the LCDA to use them in the RelQuant software.

## Results

A standard curve for the TBP amplification was generated by 1:10 dilutions of cDNA generated from INS-1E. As TBP is rather low expressed, the dilution curve covers only three orders of magnitude. From the standard curve for TBP amplification an efficiency of 1.99 was determined for the PCR (Fig. 1).

Melting curve analysis revealed amplification of unspecific byproducts exclusively in the blank (Fig. 2). These unspecific byproducts were identified by gel electrophoresis as primer dimers.

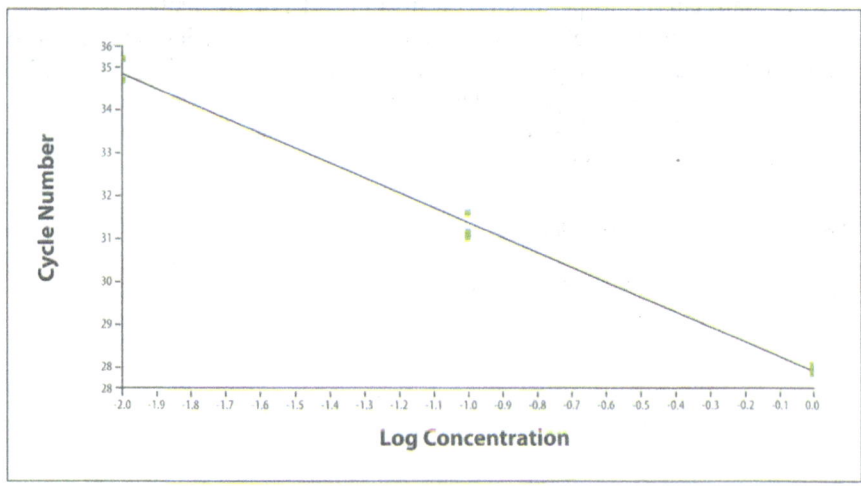

**Fig. 1.** The TBP standard curve was generated by dilution of starting material in 1:10 dilution steps. The efficiency of the reaction calculated from the slope of the curve gives a value of 1.99

As melting curve analysis of the quantification samples confirmed amplification of primer dimers exclusively in the blanks (Fig. 3) we decided to use these primers for our reference.

A standard curve of the highly expressed insulin gene was generated by 1:10 dilutions of cDNA from INS-1E material. The standard curve covers six orders of magnitude and the efficiency of the reaction was determined as 1.8 (Fig. 4).

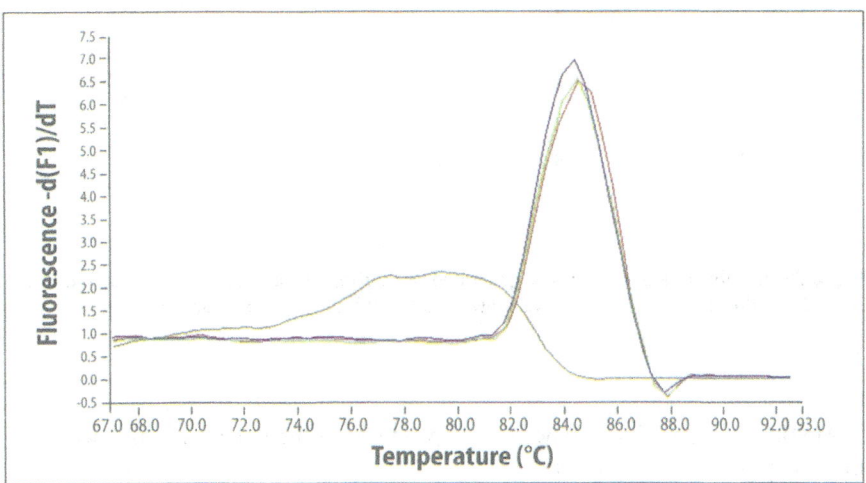

**Fig. 2.** Melting peaks of the TBP fragment generated in the standard curve with a melting temperature of 85°C and melting peak area of the byproduct in the blank that melt below 85°C and are thereby clearly distinguishable from the specific fragment

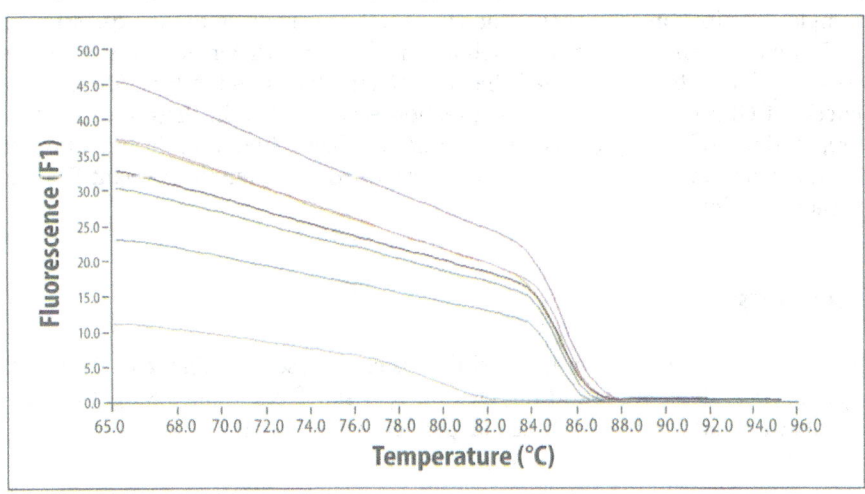

**Fig. 3.** Melting curve of the products amplified in a quantification run. Byproducts in the blank (gray) show a clearly earlier melting and are not present in any samples used for quantification

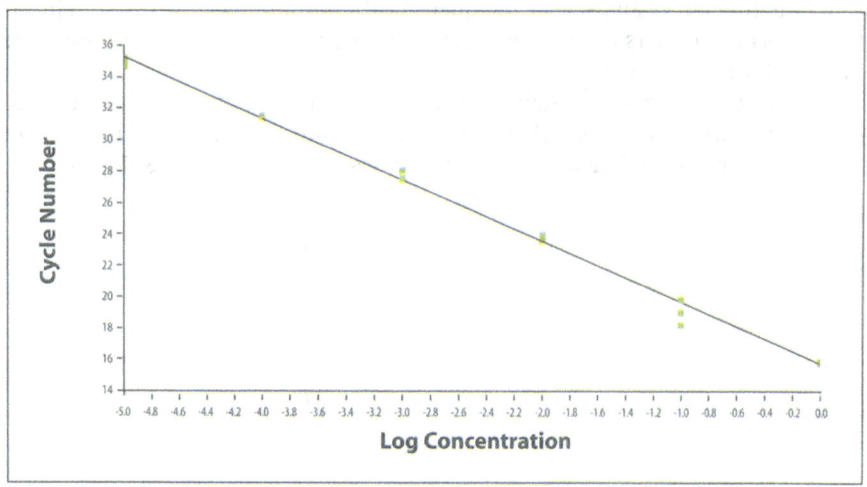

**Fig. 4.** The insulin standard curve was generated by dilution of starting material in 1:10 dilution steps. The slope of the standard curve shows that the efficiency of the reaction is 1.8

Melting curve analysis showed no byproducts in the reaction, which was confirmed by gel electrophoresis.

Both standard curves were imported in the RelQuant software to create a coefficient file that was used to correct for the different efficiency of the TBP and the insulin reaction.

For the quantification of insulin gene expression in rat islet, triplicates were run in the LightCycler for the reference and the target of controls and stimulated samples. The insulin gene expression was thereby normalized to the expression of TBP and the coefficient file corrected for the different efficiencies. This was done by the software that recognized the control samples as calibrators. Stimulatory effects of GLP-1 on insulin gene expression was analyzed in three independent preparations and analyzed using the same coefficient file. A 3-fold induction of insulin gene expression by GLP-1 was determined. An example of quantification is shown in figure 5.

## Comments

Since we had limited amounts of primary islets we used cell line material for all optimization steps. We found the method very satisfactory since we got reliable results we could confirm in three independent experiments.

```
RelQuant 1.00 - Result File

Results analysis mono color, triplicate values use coefficient file

Correction factor    : 1.000000
Multiplication factor: 1.000000

Nr   Sample info            CP    CP median   Delta CP   Ratio conc.   Normalized
                                                median                     ratio
1    control insulin        20.13    17.95     -12.18       1.15          1.00
2    Repli. of control in   17.08
3    Repli. of control in   17.95
4    control TBP            30.13    30.13
5    Repli. of control TB   34.83
6    Repli. of control TB   30.07

7    3h GLP-1 insulin       16.86    17.01     -14.11       3.83          3.32
8    Repli. of 3h GLP-1 i   18.67
9    Repli. of 3h GLP-1 i   17.01
10   3h GLP-1 TBP           31.38    31.12
11   Repli. of 3h GLP-1 T   31.12
12   Repli. of 3h GLP-1 T   28.73
```

**Fig. 5.** Example of data from the RelQuant software. Insulin expression of control samples was determined in triplicates and normalized to TBP expression. The control samples act as calibrators and the insulin expression of the stimulated samples was determined in relation to the control. CP=crossing point; CP median=median crossing point for insulin and TBP; delta CP median=difference of median CP insulin and TBP; ratio conc.= ratio of calculated concentration of gene expression

# References

1.  Buteau J, Roduit R, Susini S, Prentki M (1999) Glucagon-like peptide- 1 promotes DNA synthesis, activates phosphatidylinositol 3-kinase and increases transcription factor pancreatic and duodenal homeobox gene 1 (PDX-1) DNA bindingactivity in beta (INS-1)-cells.Diabetologia 42: 856–864
2.  Janjic D, Maechler P, Sekine N, Bartley C, Annen A, Wollheim C (1999) Free radical modulation of insulin release in INS-1 cells exposed to alloxan. Biochemical Pharmacology 57: 639–648
3.  Brunstedt J, Nielsen JH, Lernmark Å (1984) Isolation of islets from mice and rats. In: Larner J, Pohl S (eds). Methods in Diabetes Research. John Wiley and Sons, New York, vol 1:245–258
4.  Friedrichsen BN, Richter HE, Hansen JA, Rhodes CJ, Nielsen JH, Billestrup N, Moldrup A (2003) Signal transducer and activator of transcription 5 activation is sufficient to drive transcriptional induction of cyclin D2 gene and proliferation of rat pancreatic beta-cells. Mol Endocrinol 17(5):945–58

# Quantification of δRec-ψJα Signal Joint T-Cell Receptor Excision Circle DNA in Patients after Autologous and Allogeneic Stem Cell Transplantation

Juergen Loeffler[1], Holger Hebart[1], Lutz Lochmann[1], Thomas Daikeler[1], Peter Bader[2], Ralf Bauer[1], Kathrin Schmidt[1], Hermann Einsele[1]

## Introduction

Myeloablative chemotherapy followed by stem cell transplantation is often associated with a prolonged and substantial, potentially detrimental period of T-cell immunodeficiency (1). The depletion of T-cells by intensive chemotherapy may lead to an increased number of viral and fungal infections. For a complete reconstitution of immunity, the generation of de novo T-cells in the thymus with a broad T-cell receptor repertoire is essential (2). However, the thymus is gradually replaced by adipose tissue and its activity is age-dependent (3). Additional factors that influence the thymic activity are graft-versus-host disease and direct damage from chemo- and radiotherapy. Measuring thymopoietic capacity only on the basis of phenotyping naïve T-cells is of limited value as CD45RA+ T-cells may immediately convert into memory T-cells, may proliferate antigen-independently or may persist most of their life span. As an alternative approach to monitor thymic activity, the frequency of T-cell receptor excision circles among peripheral blood cells can be determined (4).

T cell receptor excision circles (TRECs) are generated during V(D)J gene recombination, a process responsible for the diversity of the T cell receptor repertoire (4). T cell receptor excision circles are circular extrachromosomal DNA fragments which are stable, do not replicate with cellular proliferation and are thus diluted with every cell division. During the TCR-α gene rearrangement and the excision of the TCR-δ locus, the δRec-ψJα recombination results in the generation of a chromosomal product and the δRec-ψJα Signal Joint TREC (5).

Here we describe a sensitive, rapid and simple real-time PCR assay based on the LightCycler technique to quantify the amount of δRec-ψJα Signal Joint TRECs among peripheral blood cells from healthy individuals and patients after autologous or allogeneic stem cell transplantation.

Data of this manuscript have been published in part elsewhere (6).

Universität Tuebingen, [1]Medizinische Klinik, Abteilung II, [2]Kinderklinik, Abteilung Hämatologie und Onkologie
Correspondence to Dr. rer. nat. Juergen Loeffler, Medizinische Klinik, Labor Prof. Dr. Einsele, In der Hals-Nasen-Ohren-Klinik, Raum 2.323, Elfriede-Aulhorn-Strasse 5, 72076 Tübingen, Germany, E-mail: juergen.loeffler@med.uni-tuebingen.de

## Materials

Equipment and
Reagents used for
the LC PCR

LightCycler™ Instrument (Roche Diagnostics, Mannheim, Germany)
Ficoll-Hypaque density gradient 1077 (Biochrom, Berlin, Germany)
Primers, Hybridization Probes (TIB MOLBIOL, Berlin, Germany)
LightCycler™- FastStart DNA Master Hyb Probes (Roche, Mannheim, Germany)
DNA Blood DNA Mini Kit (Qiagen, Hilden, Germany)

### Procedure

Patients

Patients were recruited from the University of Tuebingen, Medical Hospital (n = 13) and Childrens Hospital (n = 6). The patients received peripheral blood stem cells from an allogeneic donor (n = 13) for the treatment of acute leukemia (n = 8), chronic myelogenous leukemia (n = 3), adrenoleukodystrophy (n = 1) or Wiscott-Aldrich syndrome (n = 1). Furthermore, the number of δRec-ψJα Signal Joint TRECs in patients after autologous stem cell transplantation (n = 6) was investigated. Autologous stem cell transplantation in combination with high dose immunosuppressive therapy has been proven effective for otherwise treatment refractory autoimmune diseases. Patients analyzed were suffering from systemic scleroderma with alveolitis (n = 4), wegener's granulomatosis (n = 1) or severe psoriatic arthropathy (n = 1).

Additionally, peripheral blood from healthy volunteers (n = 68) was analyzed (median age 28.9 [range 1 - 80] years).

Sample Preparation

Peripheral blood mononuclear cells (PBMNCs) were isolated from 10 ml of EDTA-anticoagulated blood by Ficoll-Hypaque density gradient centrifugation. Genomic DNA was extracted using the QIAmp DNA Blood Mini Kit according to the protocol of the manufacturer. Purified DNA was eluted from the spin column in a concentrated form in sterile water. PCR was performed immediately or DNA was stored at -80°C until retrospective analysis.

Table 1. Oligonucleotides

| GenBank Accession #AE000661 | | | | | |
| --- | --- | --- | --- | --- | --- |
| | Position | Length | GC (%) | $T_m$ °C | Purity |
| TREC-Primers | | | | | |
| 5'- CTC TCC AAG GCA AAA TGG G | 229149–167 | 19 | 52.6 | 62.1 | 0.90 |
| 5'- GTG ACA TGG AGG GCT GAA C | 140310–292 | 18 | 55.6 | 62.8 | 0.91 |
| TREC-Probes | | | | | |
| 5'- CCA CAG GAG TGG GCA CCT TTA C-F | 229272–251 | 22 | 59.1 | 68.5 | 0.81 |
| 5'- LC Red640-AAA ACC AGA GGT GTC AG-C ATG GT- p | 229249–227 | 23 | 47.8 | 67.9 | 1.08 |

LightCycler PCR

For FastStart Taq activation, an initial 9 min at 95°C were performed before each run.

**Table 4.** Hybridization Probes Master Mix for each 20 µl reaction:

|  | Volume [µl] | [Final] |
|---|---|---|
| LightCycler™ FastStart DNA Master Hyb Probes | 2 | 1 x |
| MgCl$_2$ | 3.2 | 5 mM |
| Primers (5 µM each) | 0.1 | 0.5 µM |
| Fluorescein Probe (3µM) | 0.1 | 0.3 µM |
| LCRed640 Probe LC (3µM) | 0.2 | 0.6 µM |
| PCR grade H$_2$O | 4.4 | |
| Total volume | 10 | |

| Parameter | Value | | |
|---|---|---|---|
| Cycles | 45 | | |
| Type | Quantification | | |
|  | Segment 1 | Segment 2 | Segment 3 |
| Target temperature [°C] | 95 | 54 | 72 |
| Incubation time [s] | 3 | 15 | 25 |
| Temp. transition rate [°C/s] | 20 | 20 | 20 |
| Acquisition mode | None | Single | None |

External standardization was performed using a plasmid containing the δRec-ψJα Signal Joint TREC breakpoint. Serially diluted δRec-ψJα Signal Joint TREC DNA ranging between $2 \times 10^{10}$ copies and $2 \times 10^{1}$ copies was amplified for non-competitive external quantification in each assay.

### Results

A lower detection limit of $2 \times 10^{1}$ copies was achieved with linearity between $2 \times 10^{10}$ and $2 \times 10^{1}$ copies of plasmid DNA. Linear regression was obtained over the whole range of plasmid DNA ($r = -1.0$, error = 0.218, slope = -3.484).

*Sensitivity and Linearity*

For reproducibility experiments of the LightCycler assay, serially diluted δRec-ψJα Signal Joint TREC DNA ($2 \times 10^{6}$ to $2 \times 10^{1}$ copies) was amplified 30 times, respectively. Analyzing $2 \times 10^{6}$ standard copies, we obtained a median crossing point of 21.0 cycles (mean 20.4 cycles, standard deviation 1.05 cycles), amplifying $2 \times 10^{1}$ standard copies, a median crossing point of 36.7 cycles (mean 36.8 cycles, standard deviation 0.99 cycles) was achieved demonstrating the high reproducibility and robustness of the LightCycler assay.

*Reproducibility*

We also examined whether the number of TRECs in peripheral blood cells was correlated with age by analyzing blood specimens from healthy individuals. We achieved a median TREC count of $1.0 \times 10^{4}$ copies/$2 \times 10^{5}$ PBMNCs (mean $1.9 \times 10^{4}$ copies, range $2 \times 10^{1}$ to $2 \times 10^{6}$ copies). Twelve persons had levels of TRECs below the limit of detection by our assay and of these four were 70 years and older.

*TRECs in Healthy Persons and in Patients after Stem Cell Transplantation*

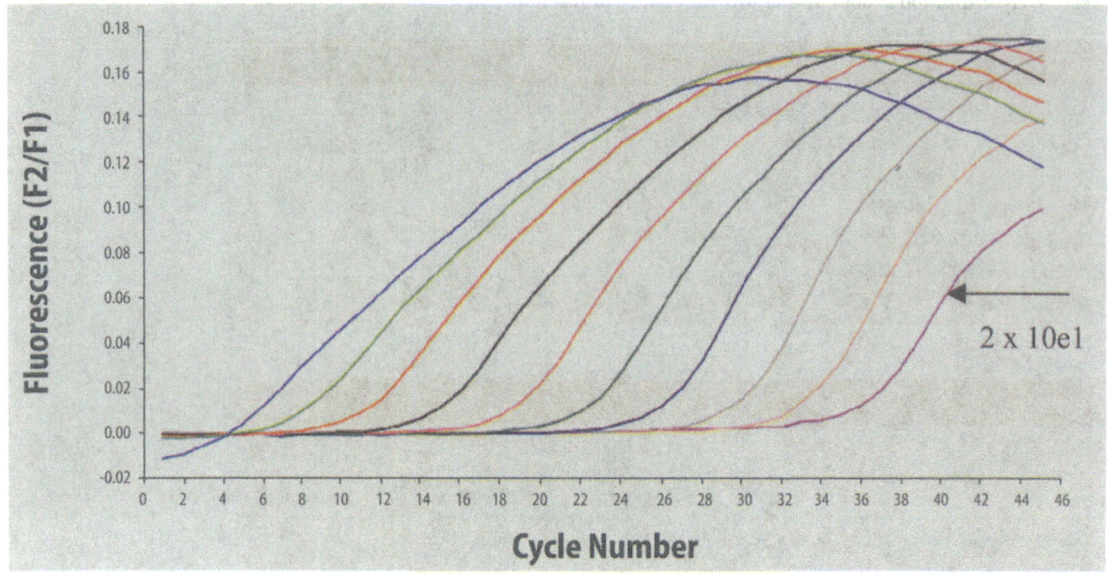

**Fig. 1.** Amplification of serially diluted plasmid δRec-ψJα Signal Joint TREC DNA (20 copies – 2 x 10⁹ copies) for external standardization (linear regression: Slope: –3.484  Error: 0.218 r: –1.00)

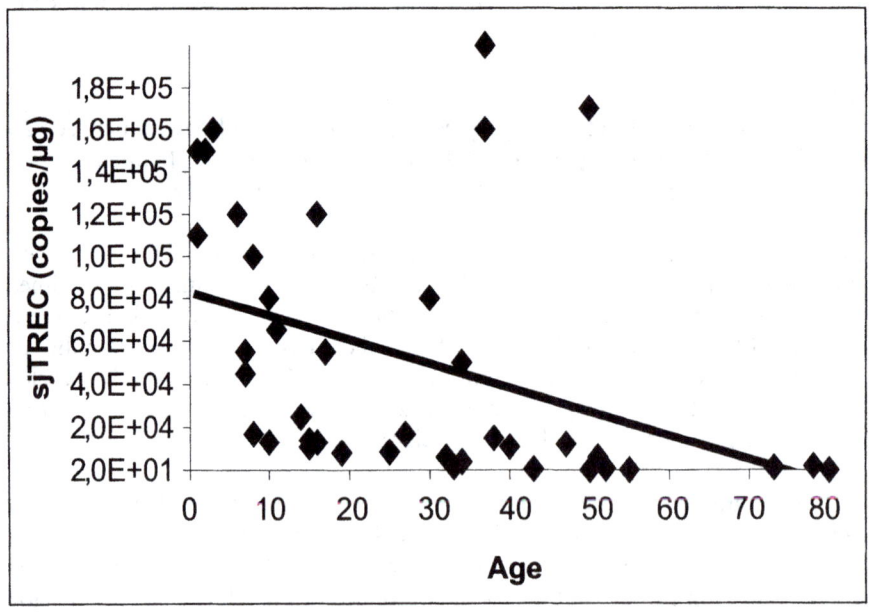

**Fig. 2.** δRec-ψJα Signal Joint TREC DNA levels (copies/μg DNA) in healthy individuals

In addition, we examined the number of δRec-ψJα Signal Joint TREC in different patient populations. After allogeneic stem cell transplantation (3, 6, 12, 18 and 24 months post transplantation), we observed in children a median count of $7.1 \times 10^4$ copies/$2 \times 10^5$ PBMNCs after 1 and $7.0 \times 10^4$ copies/$2 \times 10^5$ PBMNCs after 2 years and in adults, a median count of $3.6 \times 10^4$ copies/$2 \times 10^5$ PBMNCs after 1 and $1.1 \times 10^5$ copies/$2 \times 10^5$ PBMNCs after 2 years could be found.

In patients after autologous stem cell transplantation for therapy-refractory auto-immune diseases, we achieved a median TREC count of $3.6 \times 10^4$/µg DNA (mean $2.5 \times 10^5$/µg DNA, range $1.0 \times 10^2$ – $1.9 \times 10^6$ copies/µg DNA). δRec-ψJα Signal Joint TREC DNA was not detectable 3 and 6 months after transplantation in all patients but up to 66 months in patients after allogeneic stem cell transplantation.

## Comments

The assay has a lower detection limit of 20 copies of plasmid δRec-ψJα Signal Joint TREC DNA. Previously described assays achieved a reproducible detection limit of 500 copies of target DNA (7).

By amplifying plasmid DNA from 30 serial dilutions ($2 \times 10^6$ to $2 \times 10^1$ copies) and analyzing the crossing points of these dilution series, the LightCycler based technique provided a high reproducibility of >95%. For quantitative comparison

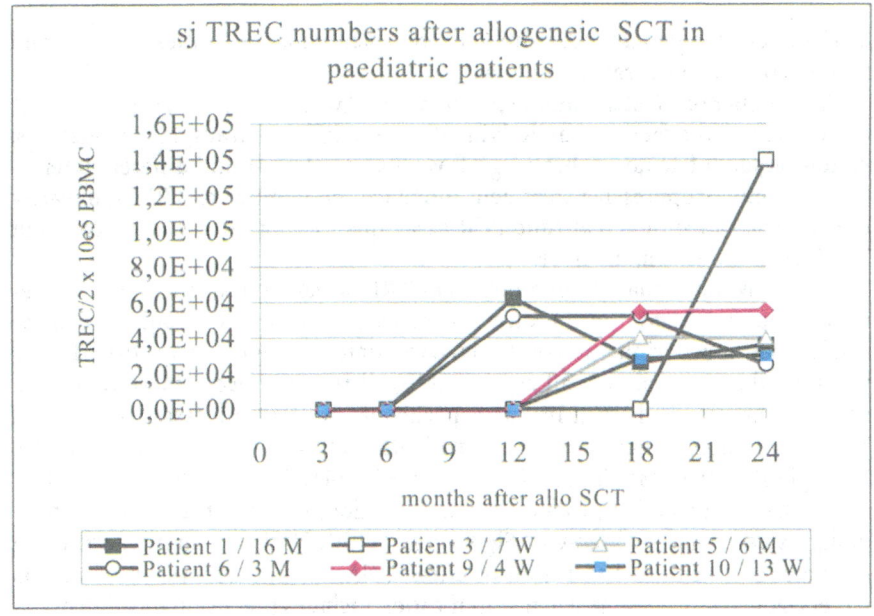

**Fig. 3.** δRec-ψJα Signal Joint TREC DNA reconstitution of 6 selected paediatric patients following allogeneic stem cell transplantation

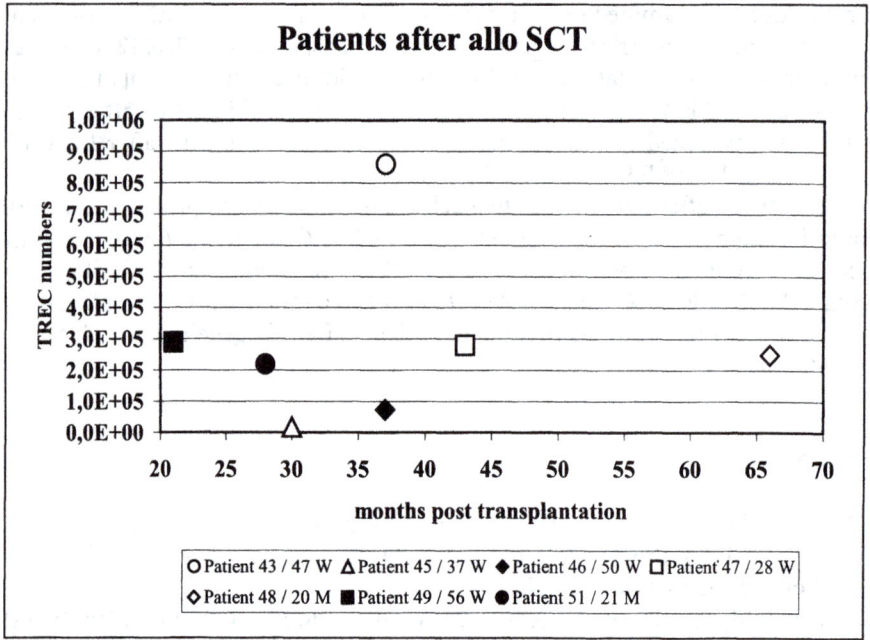

**Fig. 4.** Time after allo SCT in 9 adult patients when Rec-J – signal joint TREC DNA was detectable for the first time

of TREC levels in patients after different transplantation modalities, a robust and highly reproducible assay is essential.

The number of TRECs found in patients varied widely between 20 and $1.9 \times 10^6$ copies. Due to the fact that underlying diseases, different transplant modalities, various grades of acute or chronic graft versus host disease and different intensities of immunosuppressive treatment influence the number of TRECs, linearity over a wide range of a real-time PCR assay quantifying $\delta$Rec-$\psi$J$\alpha$ Signal Joint TREC DNA levels is mandatory.

In healthy individuals, high numbers of TRECs were found to decrease subsequently due to thymic involution with increasing age (8). We confirmed these results analyzing peripheral blood from 68 volunteers ranging in age from 1 to 80 years. Furthermore, almost all aspects of T-cell recovery occur more rapidly in children than in adults (9). In our experiments, we found a faster recovery of TREC levels after 1 year in paediatric patients compared to adults (median 7.1 x $10^4$ copies/2 x $10^5$ PBMNCs in children, 3.6 x $10^4$ copies/2 x $10^5$ PBMNCs in adults).

For successful and reproducible quantification, optimal primer and probe design is essential. As $\delta$Rec-$\psi$J$\alpha$ Signal Joint TREC DNA is a circular structure generated after end to end fusion of linear DNA, definite quantification of this target DNA is only possible when amplification includes the fusion position of the $\delta$Rec-$\psi$J$\alpha$ Signal Joint TREC DNA. Thus, in linear DNA, binding sites of the primers are located closely to both ends of the excised DNA (primer 1 at nucleotide positions 229149 – 229167, primer 2 at nucleotide positions 140293 – 140310). The length of our amplicon is 239 bp, which corresponds to the ideal length of amplicons for rapid cycling protocols.

*Acknowledgements.*
We thank O. Landt, Tibmolbiol, Berlin for his excellent technical support.
We thank D. Douek, NIH Vaccine Research Center, Bethesda, USA, for the plasmid containing the δRec-ψJα Signal Joint TREC breakpoint.

# References

1. Keever, C.A., Small, T.N., Flomenberg, N., Heller G., Pekle K., Black P., Pecora A., Gillio A., Kernan N.A., O`Reilly R.J. (1989) Immune reconstitution following bone marrow transplantation: comparison of recipients of T-cell depleted marrow with recipients of conventional marrow grafts. *Blood* 73: 1340–1350

2. Heitger A., Greinix, H., Mannhalter, C., Mayerl D., Kern H., Eder J., Fink F.M., Nierderwieser D., Panzer-Grumayer E.R. (2000) Requirement of residual thymus to restore normal T-cell subsets after human allogeneic bone marrow transplantation. *Transplantation* 69: 2366–2373

3. Douek, D.C., McFarland, R.D., Keiser, P.H., Gage, E.A., Massey, J.M., Haynes, B.F., Polis, M.A., Haase, A.T., Feinberg, M.B., Sullivan, J., Jamieson, B.D., Zack, J.A., Picker, L.J., Koup, R.A. (1998) Changes in thymic function with age and during the treatment of HIV infection. *Nature* 396: 690–695

4. Douek, D.C., Vescio, R.A., Betts, M.R., Brenchley, J.M., Hill, B.J., Zhang, L., Berenson, J.R., Collins, R.H., Koup, R.A. (2000) Assessment of thymic output in adults after haematopoietic stem cell transplantation and prediction of T-cell reconstitution. *Lancet* 355: 1875–1881

5. Verschuren M.C.M., Wolvers-Tettero I.L.M., Breit T.M., Noordzij J., van Wering E.R., van Dongen J.J.M. (1997) Preferential Rearrangements of the T cell Receptor-δ-Deleting elements in human T cells. *J Immunol* 158: 1208–1216

6. Loeffler, J., Bauer, R., Hebart H., Douek D., Rauser G., Bader P., Einsele H. (2002) Quantification of T-Cell Receptor Excision Circle DNA by Fluorescence Resonance Energy Transfer and the LightCycler system. *J Immunol Meth* 261: 167–175

7. Al-Harthi, L., Marchetti, G., Steffens, C.M., Poulin, J.F., Sekaly R.P., Landay A. (2000) Detection of T cell receptor circles (TRECs) as biomarkers for de novo T cell synthesis using a quantitative polymerase chain reaction enzyme linked immunosorbent assay (PCR-ELISA). *J. Immunol. Meth.* 237: 187–197

8. Hochberg, E.P., Chillemi, A.C., Wu, C.J., Neuberg, D., Canning, C., Hartman, K., Alyea, E.P., Soiffer, R.J., Kalams, S.A., Ritz, J. (2001) Quantitation of T-cell neogenesis in vivo after allogeneic bone marrow transplantation in adults. *Blood* 98: 1116–1121

9. Patel D.D., Gooding, M.E., Parrott, R.E., Curtis, K.M., Haynes, B.F., Buckley, R.H. (2000) Thymic function after hematopoietic stem cell transplantation for the treatment of severe combined immunodeficiency. *N Engl J Med* 342: 1325–1332

# Expression Analysis of Mitochondrial Components in a Variety of Plant Species Using Real-Time Quantitative PCR

Katharine A. Howell, Ryan Lister, James Whelan*

## Introduction

The endosymbiotic origin of mitochondria has resulted in the existence of a genetic system distinct from that associated with the nucleus. However, the mitochondrial genome only encodes a small proportion of the many hundreds of proteins present in the mitochondria indicating that, over time, most of the genes originally present in the organellar genome have been transferred to the nucleus [1]. More specifically, many of the large multi-subunit complexes present in the mitochondria are composed of proteins encoded in both the nuclear and mitochondrial genomes. Therefore, the process of mitochondrial biogenesis requires the coordination of nuclear and mitochondrial gene expression for the generation of functional complexes with specific subunit stoichiometries.

In plants, approximately 95% of mitochondrial proteins are nuclear encoded. As synthesis of these proteins occurs in the cytosol, proteins need to be transported into the mitochondria so they can perform their appropriate functions. The protein-import apparatus, which plays a central role in mitochondrial biogenesis, mediates this process [2].

To understand the process of mitochondrial biogenesis, we examined the expression of various mitochondrial components in several plant species. We successfully measured transcript levels of electron transport chain complex subunits, tricarboxylic acid cycle enzyme subunits, mitochondrial import components, ribosomal proteins, mitochondrial division proteins, carrier proteins, the uncoupling protein (Ucp), and the alternative oxidase (Aox), in a variety of plants including rice, *Arabidopsis* [3], soybean [4] and mango [5]. Both nuclear and mitochondrial encoded genes were investigated, which allowed us to understand the regulation of mitochondrial biogenesis at the gene expression level. Spatial and temporal patterns of expression were also examined to study mitochondrial components over development and in different tissues.

Quantitative real-time PCR, using the LightCycler system, was used to perform this investigation of expression levels. The LightCycler is an ideal system for this

---

* James Whelan, Plant Molecular Biology Group, Biochemistry and Molecular Biology, School of Biomedical and Chemical Sciences, University of Western Australia, 35 Stirling Highway, Crawley, 6009, WA, Australia, E-mail: seamus@cyllene.uwa.edu.au

study because it is reproducible, quantitative, has a relatively high-throughput and only requires small amounts of starting material when compared to other methods. Here we describe examples from studies that demonstrate the approaches we have used with the LightCycler system for analyzing expression levels of mitochondrial components in a variety of plant species.

### Materials

**Equipment**

LightCycler® instrument (Roche, Sydney, Australia)
LightCycler® software, version 3.5 (Roche, Sydney, Australia)
LightCycler® capillaries, centrifuge adapters, and cooling block (Roche, Sydney, Australia)

**Reagents**

RNeasy® Plant Mini Kit (Qiagen, Clifton Hill, Australia)
RNase-Free DNase Set (Qiagen, Clifton Hill, Australia)
DNA-*free*™ kit (Ambion, Austin, Texas, USA)
Expand reverse transcriptase (Roche, Sydney, Australia)
Random primer p(dN)$_6$ (Roche, Sydney, Australia)
Oligo-p(dT)$_{15}$ primer (Roche, Sydney, Australia)
LightCycler-FastStart DNA Master SYBR Green I (Roche, Sydney, Australia)
Bovine serum albumin, acetylated, 10 mg/ml (Promega, Annandale, Australia)
Oligonucleotide primers (Sigma, Castle Hill, Australia)
Expand High Fidelity PCR System (Roche, Sydney, Australia)
TOPO TA Cloning® kit (Invitrogen, Sydney, Australia)
PicoGreen dsDNA Quantitation Kit (Molecular Probes, Eugene, USA)
QIAquick® PCR Purification Kit (Qiagen, Clifton Hill, Australia)
QIAquick® Gel Extraction Kit (Qiagen, Clifton Hill, Australia)

## Procedure

**Preparation of cDNA**

Plant tissue was ground to a fine powder, under liquid nitrogen, using a mortar and pestle. Total RNA was isolated from rice embryos; *Arabidopsis* cotyledons, roots, leaves and floral tissue; and soybean cotyledons using the RNeasy Plant Mini Protocol. Total RNA was isolated from mango fruit using a modified hot-borate procedure developed for tissues rich in polysaccharides [6]. Three independent RNA preparations were performed for each tissue and/or developmental stage. Genomic DNA was removed from rice and *Arabidopsis* RNA using the RNase-Free DNase Set and the DNA-*free* kit. Genomic DNA was removed from soybean and mango RNA using DNase I (Roche). The concentration and integrity of the RNA was determined spectrophotometrically and by agarose gel electrophoresis respectively.

cDNA was prepared from each sample using 1 µg of total RNA, Expand Reverse Transcriptase and random priming with p(dN)$_6$ (for rice and *Arabidopsis* samples) or oligo-p(dT)$_{15}$ (for mango and soybean samples). For each RNA sample, three reverse transcription reactions were performed (two with the reverse transcriptase enzyme present and one without, to check for genomic DNA contamination). Samples of cDNA were then purified using the QIAquick PCR Purification Kit. For real-time PCR analysis, cDNA samples were diluted 1 in 10 with water and with a final concentration of 0.008% (w/v) BSA.

For each gene of interest, a complete or partial fragment was amplified from cDNA or genomic DNA using the Expand High Fidelity PCR system and a standard thermocycler. PCR products of the desired size were extracted from a gel slice using the QIAquick Gel Extraction Kit, or directly purified from the PCR reaction using the QIAquick PCR Purification Kit. The purified PCR product was then cloned using the TOPO TA Cloning kit and the identity of the insert confirmed by restriction digest analysis and sequencing. A linear standard for real-time PCR was then prepared by amplifying the insert using the cloning primers and a standard thermocycler. Purification of the PCR product was performed using the QIAquick PCR Purification Kit and the DNA concentration of this external standard was determined using the PicoGreen dsDNA Quantitation Kit. External standards were then diluted to 0.01 $fmol/\mu l$ to be used as a template for LightCycler PCR.

**Preparation of External Standard**

Primers for LightCycler PCR were selected and designed according to the recommendations outlined in the LightCycler Operator's Manual. Where possible, primers were designed spanning introns (when genomic sequence was available) and for members of gene families, areas of low identity were chosen. After resuspension, primers were diluted to 20 $\mu M$ in PCR grade water with a final concentration of 0.008% (w/v) BSA. Primer pairs were optimized using the external standard prepared for each gene of interest and the optimum $Mg^{2+}$ and primer concentration determined to maximize amplification efficiency and specificity.

**Oligonucleotides**

LightCycler reactions were performed in a total volume of 10 $\mu l$ consisting of 9 $\mu l$ of Master Mix (see below) and 1 $\mu l$ of template (diluted cDNA sample, external standard or PCR grade water as a negative control).

**LightCycler PCR**

| | Volume [µl] | [Final] |
| --- | --- | --- |
| LightCycler-FastStart DNA Master SYBR Green I | 1.0 | 1 × |
| MgCl₂ (25 mM) | 0.8/1.2/1.6 | 3.0/4.0/5.0 mM |
| Primers (20 µM each) | 0.15/0.25/0.35 each | 0.3/0.5/0.7 µM each |
| BSA (0.08% (w/v)) | 1.0 | 0.008% (w/v) |
| H₂O (PCR grade) | to a total volume of 9.0 | – |

**Table 1.** Oligonucleotides

| Rice cox2 subunit of cytochrome c oxidase (GenBank Accession no. X01088) | | | | |
|---|---|---|---|---|
| Primer sequences | Length (nt) | GC (%) | $T_m$ (°C) | Amplicon (bp) |
| CGAGCAAACTAATCCAATCC | 20 | 45.0 | 60.8 | 205 |
| GTCCGAATACTCATAAGTCC | 20 | 45.0 | 59.0 | |
| Rice cox5b subunit of cytochrome c oxidase (GenBank Accession no. D85381) | | | | |
| GCGGTTCGACATGGATCC | 18 | 61.1 | 65.0 | 167 |
| GGCACTCATGCGGTTCATC | 19 | 57.9 | 65.8 | |
| *Arabidopsis* TIM17–1 (TAIR Accession no. At1g20350) | | | | |
| CGTTCAAGCTTTGAGAATG | 19 | 42.1 | 58.9 | 278 |
| GCTCGTTATGCGCAGTAC | 18 | 55.6 | 63.0 | |
| *Arabidopsis* TIM17–2 (TAIR Accession no. At2g37410) | | | | |
| GTGAGCATGAACGCACCTCG | 20 | 60 | 66.2 | 301 |
| CAGGCATTCCTTGCATTCCAGG | 22 | 54.5 | 66.2 | |
| *Arabidopsis* TIM17–3 (TAIR Accession no. At5g11690) | | | | |
| CTAAGGAACATGGCCTATACC | 21 | 47.6 | 61.5 | 290 |
| CAGAACTCCACCTGTAGC | 18 | 55.6 | 61.1 | |
| Soybean $F_A$d subunit of ATP synthase (GenBank Accession no. X79058) | | | | |
| GGAGCATGCTATTCCAGTTCG | 21 | 52.4 | 65.3 | 232 |
| GGACTCCGACAATAAGCTTGC | 21 | 52.4 | 65.5 | |
| Soybean $F_1\beta$ subunit of ATP synthase (GenBank Accession no. AF529298) | | | | |
| GGAGCTGCCCTAAGTGTTCC | 20 | 60.0 | 64.8 | 209 |
| CCTCCACGACGATAAGGAGC | 20 | 60.0 | 64.3 | |
| Mango Aox1a (GenBank Accession no. AF329895) | | | | |
| GCTGTTACACTGCAAGTCAC | 20 | 50.0 | 62.4 | 180 |
| AAGTAGGCGTTGAAGAACACG | 21 | 47.6 | 63.9 | |
| Mango Aox1b (GenBank Accession no. AF329896) | | | | |
| GGTGGGCGGCATGCTGTTACACTGC | 25 | 64.0 | 75.2 | 191 |
| GAAGAATACTCCTTGGACTGCC | 22 | 50.0 | 63.4 | |
| Mango Aox2 (GenBank Accession no. AF329898) | | | | |
| TTGCTGCATCTGAAGTCTCTCC | 22 | 50.0 | 65.3 | 168 |
| AAAGAAGACTCCCTGCACAGCA | 22 | 50.0 | 67.2 | |

The following PCR protocol was used for amplification and melting curve analysis:
- Denaturation for 10 min at 95°C
- Amplification

| Parameter | Value | | |
|---|---|---|---|
| Cycles | 50 | | |
| Type | Quantification | | |
| | Segment 1 | Segment 2 | Segment 3 |
| Target temperature [°C] | 95 | 85 | 72 |
| Incubation time [s] | 15 | 5 | 10 |
| Temperature transition rate [°C/s] | 20 | 20 | 20 |
| Secondary target temperature [°C] | – | 50 | – |
| Step size [°C] | – | 2.0 | – |
| Acquisition mode | None | None | Single |

- Melting Curve Analysis

| Parameter | Value | | |
|---|---|---|---|
| Cycles | 1 | | |
| Type | Melting Curve Analysis | | |
| | Segment 1 | Segment 2 | Segment 3 |
| Target temperature [°C] | 95 | 70 | 95 |
| Incubation time [s] | 0 | 30 | 0 |
| Temperature transition rate [°C/s] | 20 | 20 | 0.1 |
| Acquisition mode | None | None | Continuous |

- Cooling for 30 s at 40°C

Levels of transcripts were determined by generating a standard curve using at least four 10-fold dilutions of the standard (in duplicate) and using the second derivative maximum method to determine transcript abundance in the cDNA samples. Each cDNA sample was analyzed in duplicate. PCR specificity was confirmed by melting curve analysis and agarose gel electrophoresis.

## Results

Transcript levels of mitochondrial components in rice embryos were analyzed from 0 to 48 h post-imbibition (8 time points in total). Imbibition refers to the process of water absorption by the dry seed. Real-time quantitative PCR was an obvious choice for this study considering the very low amounts of available sample material (rice embryos). cDNA used in this study was prepared using random

Rice

priming so measurements of nuclear and mitochondrial encoded components could be performed on the same cDNA pool (mitochondrial transcripts usually lack a poly[A] tail). Prior to transcript analysis, controls lacking reverse transcriptase were used to check for genomic DNA contamination. External standards were chosen such that all cDNA samples were amplified within the range of the standard curve. Quantification with an external standard curve (LightCycler software version 3.5) was used so all cDNA samples could be analyzed for each gene in two runs.

Over 30 genes were examined as part of this study, but here we present results of transcript levels for *cox2*, a mitochondrial encoded subunit, and *cox5b*, a nuclear encoded subunit, of the cytochrome *c* oxidase complex of the electron transport chain. A typical amplification graph, standard curve, and derivative melting curve plot are shown in Figure 1 for *cox2* analysis. Both standards and cDNA samples showed an exponential curve (Figure 1a) typical of real-time PCR while the negative control showed no increase in fluorescence. The standard curve derived from this data (Figure 1b) indicates an amplification efficiency of 1.95. Derivative melting curve data (Figure 1c) shows that a specific product is generated when both external standards and cDNA samples are used as a template. Agarose gel electrophoresis of PCR products confirmed that products were the expected size of 205 bp (data not shown).

By using an external standard, absolute and relative levels of *cox2* and *cox5b* transcripts could be estimated. Interestingly, a similar developmental trend was found when relative levels of both subunits were compared (Figure 2a). However, the absolute levels differed greatly, with transcript levels of *cox2* (mitochondrial encoded) hundreds of times higher than levels of *cox5b* (nuclear encoded; Figure 2b). This is not surprising since the copy number of the mitochondrial genome is higher than that of the nuclear genome since there are many copies of the mitochondrial genome per mitochondria and many mitochondria per cell.

Standard errors were calculated for each data point. It was found that most variability was associated with the different RNA samples rather than the different reverse transcription reactions or duplicate LightCycler runs. Despite this, variation was still minimal and statistically significant changes in transcript levels were detected.

**Arabidopsis**

Expression analysis of various transmembrane import components was performed on different *Arabidopsis* tissues and at different developmental stages [3]. In *Arabidopsis*, *Tim17*, a component of the translocase of the inner mitochondrial membrane, is encoded by a small gene family consisting of three isoforms: *Tim17-1*, *Tim17-2* and *Tim17-3*. To examine levels of these different transcripts it was necessary to check for primer specificity so message abundance of each isoform could be determined independently. Primer pairs were designed and then checked for cross-reactivity using the different *Tim17* standards and it was found that each primer pair was specific for its corresponding isoform.

Using these primers, transcript levels of each *Tim17* isoform were determined in different *Arabidopsis* tissues including cotyledons, leaves, roots, and floral tissue (Figure 3a). In addition, a more comprehensive study over cotyledon devel-

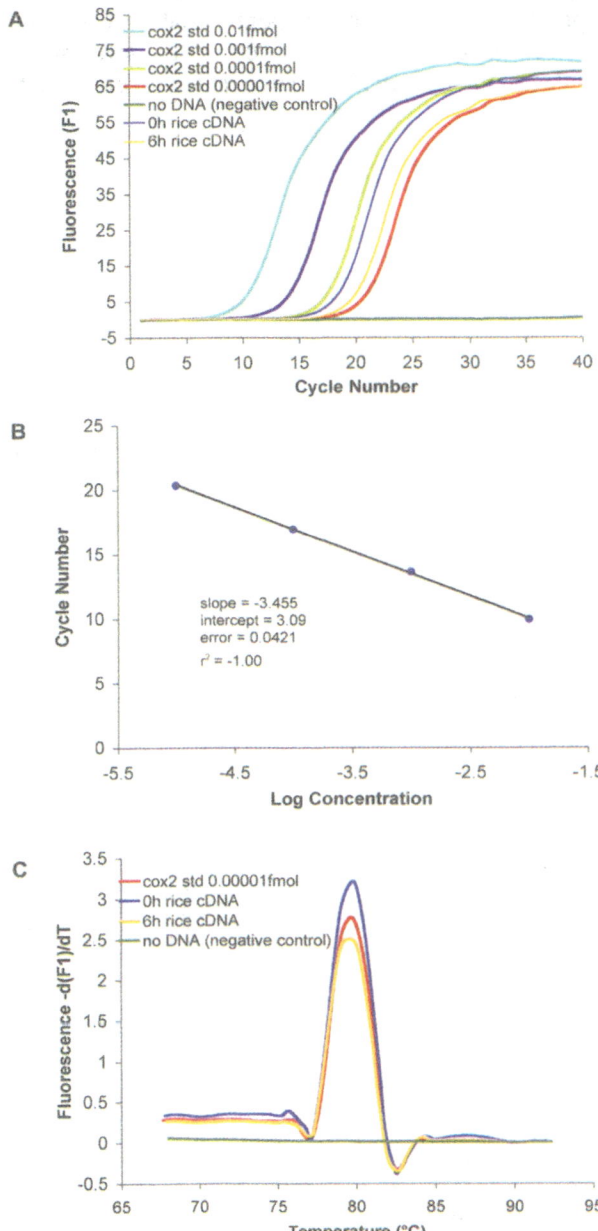

**Fig. 1.** Typical data generated on the LightCycler. (a) Plot of fluorescence versus cycle number using *cox2* primers with the *cox2* standard and rice embryo cDNA samples (0 h and 6 h post-imbibition). (b) Standard curve generated from the *cox2* external standards showing cycle number plotted against log concentration (*f*mol). (c) Melting peak data for the *cox2* external standard, rice embryo cDNA samples and the negative control

opment was performed (Figure 3b). It was found the *Tim17* isoforms were differentially expressed with respect to both spatial and temporal variation. For example, *Tim17-1* was not expressed in roots, while during cotyledon development *Tim17-3* showed a different developmental trend in transcript levels compared to *Tim17-1* and *Tim17-2*.

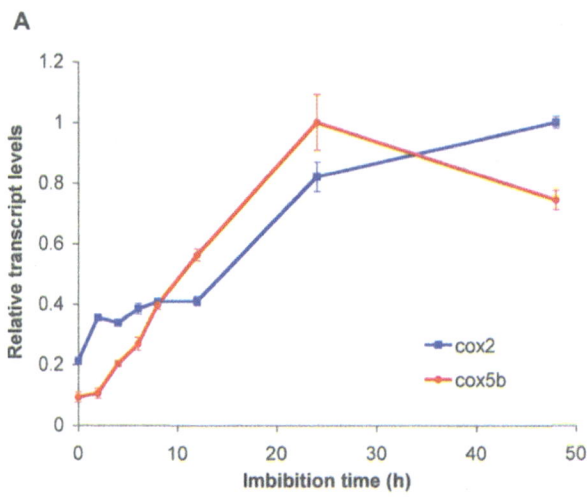

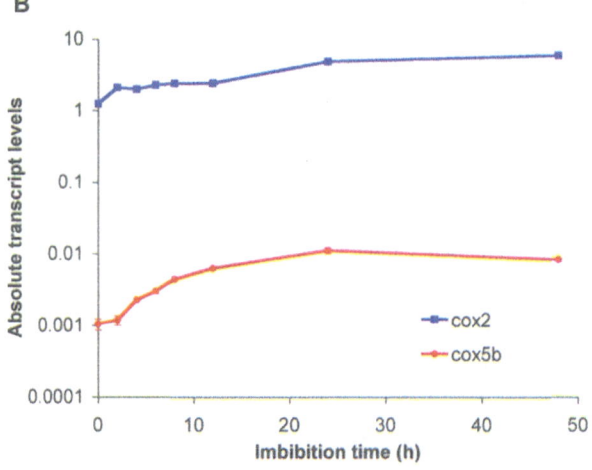

**Fig. 2.** Transcript levels of mitochondrial encoded (*cox2*) and nuclear encoded (*cox5b*) subunits of the cytochrome *c* oxidase complex in rice embryos over 48 h post-imbibition. (a) Relative transcript levels of *cox2* and *cox5b*. The developmental stage with the highest transcript level was set to 1 and the other stages normalized to it. (b) Absolute transcript levels ($10^{-19}$ mol of transcripts/$\mu$g of total RNA) of *cox2* and *cox5b* estimated using the external standard of known concentration

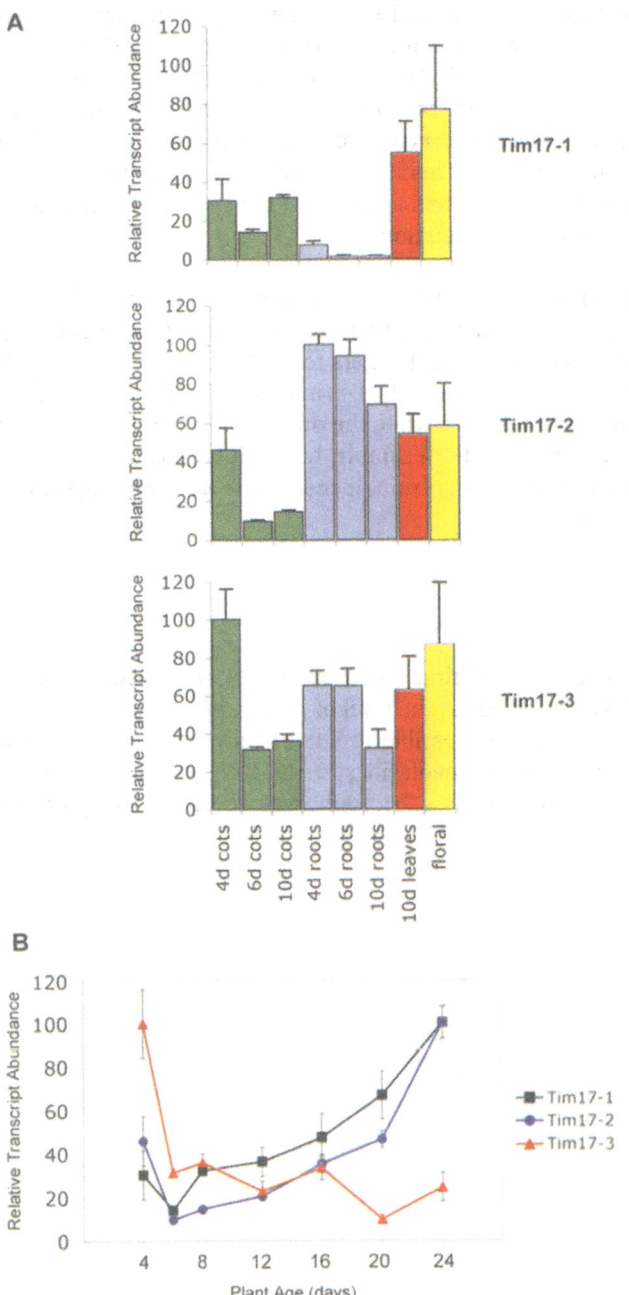

**Fig. 3.** Transcript levels of *Tim17–1, Tim17–2,* and *Tim17–3* in different *Arabidopsis* tissues and over development. (a) Relative transcript levels in *Arabidopsis* cotyledons, roots, leaves, and floral tissue. (b) Relative transcript levels in cotyledons over development

**Soybean**    Respiratory gene expression was examined in soybean cotyledons during post-germinative development [4]. Figure 4 shows relative transcript levels determined using the LightCycler for the $F_1\beta$ and $F_A d$ subunits (both nuclear-encoded) of the ATP synthase complex of the electron transport chain. Interestingly, similar patterns of transcript abundance were observed, except at day 5 and 7 when levels of $F_1\beta$ transcripts peaked dramatically. This increase in transcript abundance was not associated with an increase in protein levels (data not shown, [4]) indicating translational or post-translational regulation.

**Mango**    The expression of alternative oxidase (*Aox*) and uncoupling protein (*Ucp*) was examined during mango fruit ripening [5]. Four Aox isoforms were identified in mango - *Aox1a, Aox1b, Aox1c* and *Aox2*. Primers for real-time quantitative PCR were designed and tested for cross-reactivity. During the five stages of ripening, transcripts of *Aox1c* were not detected, while the other isoforms were expressed in a differential manner (Figure 5). For example, transcript levels of *Aox1a* and *Aox1b* started low and increased as the fruit ripened while *Aox2* message abundance decreased with ripening.

## Comments

Using the LightCycler system for real-time quantitative PCR, we have been able to investigate mitochondrial biogenesis at the level of gene expression in a number of plant species. We have found it possible to successfully measure transcript abundance of mitochondrial components in systems that do not allow large amounts of material to be collected (e.g., rice embryos and *Arabidopsis* tissue)

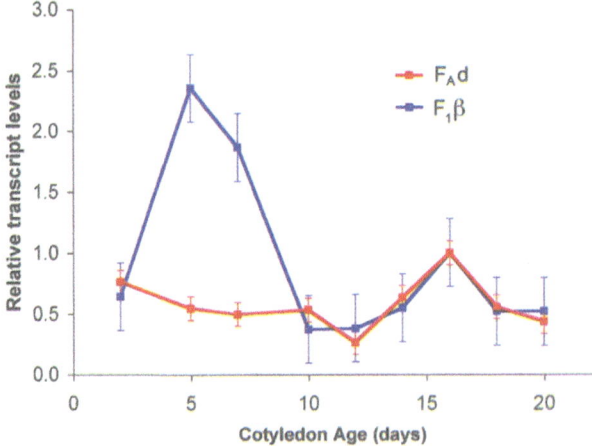

**Fig. 4.** Relative transcript levels of the $F_1\beta$ and $F_A d$ subunits of the ATP synthase complex in soybean cotyledons over post-germinative development. Transcript abundance at the 16 d stage was set to 1 and the other stages normalized to it

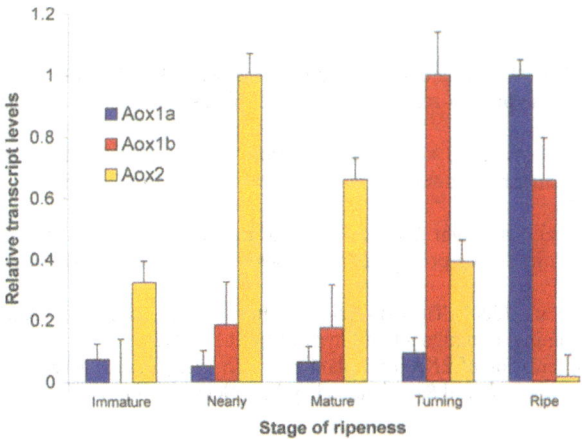

**Fig. 5.** Relative transcript levels of the Aox isoforms in mango fruit at different stages of ripeness. The developmental stage with the highest transcript level was set to 1 and the other stages normalized to it

and in tissues that are usually difficult to work with (e.g., mango fruit, which is high in polysaccharides). We found that it was best to treat the RNA isolated from these tissues twice with DNase I to completely remove contaminating genomic DNA as the sensitivity of the LightCycler resulted in the detection of genomic contamination in about 30% of those samples only treated once with DNase I. Two-step RT-PCR was performed to allow transcript levels of a number of genes to be determined from a common cDNA pool.

As recommended by Roche, we used BSA (to a final concentration of 0.008% [w/v]) in primer dilutions, cDNA dilutions and LightCycler reactions to stabilize dilutions and improve reproducibility. A generic touch-down, hot-start program was formulated for use with the LightCycler and was successful for a range of primer pairs in terms of generating specific products in real-time PCR. Therefore, a standard approach for quantitative real-time PCR has been established including: (1) a generic touch-down PCR protocol, (2) preparing cDNA with random primers so all transcripts are represented and (3) using SYBR Green I instead of labeled probes. With this standard approach, measuring transcript levels of other genes in previously prepared cDNA samples is very easy and the time required to set up new systems is also minimized.

Future directions in real-time PCR using the LightCycler include moving into other systems such as tobacco and *Arabidopsis* cell cultures. Apart from quantitating transcript levels, we are also using the LightCycler system to estimate genome copy number (nuclear, mitochondrial, and chloroplastic). In addition, as we move towards more high-throughput systems such as microarrays, real-time PCR will be useful in terms of confirming changes in gene expression determined from the microarray data. Using cluster analysis of transcript profiles, followed by promotor analysis, identification of potentially important promotor elements may be possible.

# References

1. Gray MW, Burger G, Lang BF (1999) Mitochondrial evolution. Science 283: 1476–1481
2. Braun HP, Scmitz UK (1999) The protein-import apparatus of plant mitochondria. Planta 209: 267–274
3. Murcha MW, Lister R, Ho AYY, Whelan J (2003) Identification, expression and import of components 17 and 23 of the inner mitochondrial membrane translocase from Arabidopsis. Plant Physiol 131: 1737–1747
4. Daley DO, Considine MJ, Howell KA, Millar AH, Day DA, Whelan J (2003) Respiratory gene expression in soybean cotyledons during post-germinative development. Plant Mol Biol 51: 745–755
5. Considine MJ, Daley DO, Whelan J (2001) The expression of alternative oxidase and uncoupling protein during fruit ripening in mango. Plant Physiol 126: 1619–1629
6. Wilkins TA, Smart LB (1996) Isolation of RNA from plant tissue. In Kreig PA (ed) A Laboratory Guide to RNA: Isolation, Analysis and Synthesis, Wiley-Liss, New York, pp21–42

# Quantification of Ikaros Splice Variants by Real-Time PCR

Elli Veistinen*, Kalle-Pekka Nera, Jukka Alinikula, Olli Lassila

## Introduction

The Ikaros family is an important group of transcription factors essential in the normal development of hematopoietic lineages [1, 2, 3]. They function as transcriptional regulators through chromatin remodeling [4] and are active only as dimers [1]. Ikaros family members – Ikaros, Aiolos, and Helios – are DNA-binding, zinc-finger proteins. They all generate a large pattern of splice variants [1, 5, 6, 7, 8]. Cloned Ikaros isoforms are shown in Figure 1. Some of the splice variants lack one or more of the four N-terminal zinc fingers and their DNA binding affinity is altered or totally lost. Isoforms that lack the DNA-binding capacity are thought to have a dominant negative function because they prevent the binding of the dimer to DNA. Ikaros family proteins function as part of a multicomponent chromatin remodeling complex [9]. Splice variants missing one or more exons might selectively lack sites for protein-protein interactions or regulation of the complex, and therefore, have altered function.

The function of exon 6 of Ikaros gene is not known. However, it contains a conserved area of 15 amino acids that is shared with other Ikaros family members. In addition, a 10 amino acid deletion in exon 6 has been reported in patients with acute lymphoblastic leukaemia [10, 11]. Deletion partly removes the conserved area. We wanted to measure the relative expression of Ikaros isoforms lacking exon 6 in the development of chicken embryos to reveal any alterations in the expression pattern, and characterize the function of exon 6. To address the question we used a hybridization probe-based method described earlier for Aiolos isoforms [12]. We chose real-time PCR because it is a fast and reproducible high-throughput method. We changed the earlier described method slightly to a dual color application, where all the probes are set to the same reaction.

## Materials

LightCycler® instrument (Roche Diagnostics, Mannheim, Germany)    Equipment

---

* Elli Veistinen, Turku Graduate School of Biomedical Sciences and Department of Medical Microbiology, University of Turku, Kiinamyllynkatu 13, 20520 Turku, Finland
E-mail: elli.veistinen@utu.fi

Reagents    LightCycler FastStart DNA Master Hybridization Probes (Roche Molecular Bio-
chemicals, Mannheim, Germany)
Amplification primers (TibMolbiol, Berlin, Germany)
Fluorescein-labeled probes (TibMolbiol, Berlin, Germany)
TRIzol® reagent (Life Technologies, NY, USA)
First Strand cDNA Synthesis Kit for RT-PCR (AMV) (Roche Molecular Biochem-
icals, Mannheim, Germany)

## Procedure

Sample Preparation    RNA was isolated from chicken cell lines DT40, RP-13, Cu-17, Cu-32 and Cu-36,
and also from day 2 and 5 whole chicken embryonic cells [13, 14, 15, 16]. Total
RNA was isolated with TRIzol reagent according to the manufacturer's instructions. 1 μg of total RNA was used for a reverse transcription reaction (total volume 20μl) with AMV reverse transcriptase and oligo-p(dT) primers (First Strand
cDNA Synthesis Kit for RT-PCR). Negative controls were made by replacing the
reverse transcriptase with water. The reactions were incubated at 25C for 10 min
and then at 42C for 1 h, and finally, at 95C for 5 min.

Avian Ikaros splice variants were TA-cloned using the pGEM®-T Easy Vector
system (Promega Corporation) and sequenced as previously described [6]. Splice
variant clones corresponding to chicken Ikaros isoforms Ik-1 and Ik-1A were
used to detect the probe pairs' specificity. Dilutions of $1 \times 10^{-3}$ and various mixes
of the diluted isoforms were used as a template in a LightCycler run.

Oligonucleotides    The primers and hybridization probes listed in Table 1 were chosen for the chick-
en Ikaros gene. Probe pair 1 measures the total amount of all Ikaros isoforms and
probe pair 2 measures the amount of isoforms, where exon 6 is missing (Figure 2).
The primers and probes were designed by TibMolbiol (Berlin, Germany). Probe
pair 1 binds to exon 7, which is present in all isoforms. The donor probe of the
second pair is at the end of exon 5, and the acceptor probe is at the beginning of

**Table 1.** Oligonucleotides

| Gallus Gallus Ikaros (GenBank Accession #Y11833) | | | |
|---|---|---|---|
| | Length | GC (%) | $T_m$ (°C) |
| Primers | | | |
| TGGTCGCAGCTATAAGCAG | 19 | 52.6 | 54.1 |
| GCACCACTTCAGAACCAAC | 19 | 52.6 | 53.6 |
| Probe pair 1 | | | |
| GGAGAACGAGATAATGCAGACCCA –F | 24 | 50.0 | 60.3 |
| LCRED705-GTCATAGACCAGGCCATTAACAATG-P | 25 | 44.0 | 59.8 |
| Probe pair 2 | | | |
| ACCATGAGTATCTCAAGCAATCTTTATTCA-F | 30 | 33.3 | 59.1 |
| LCRed640-GAGAAGTGTCTATCTGATCTTCCATATGATG-P | 31 | 38.7 | 58.9 |

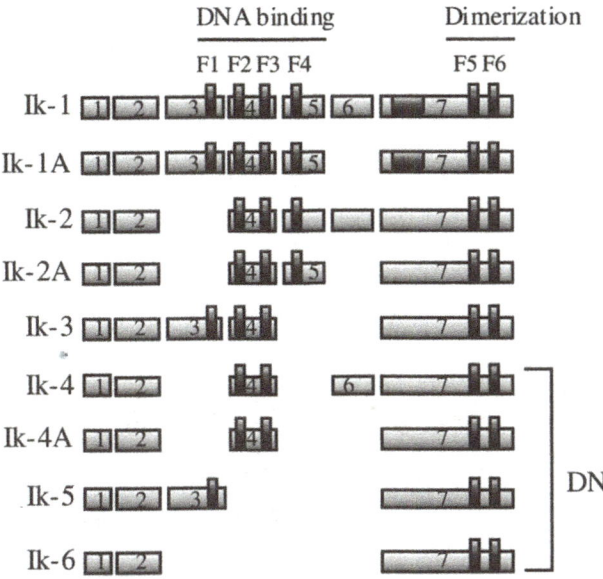

**Fig. 1.** Cloned Ikaros isoforms. Dominant Negative = DN. F1-F6 are zinc fingers [20, 25, 26, 27, 28]

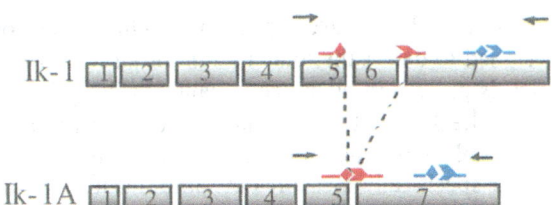

**Fig. 2.** The hybridization probes on the chicken Ikaros gene. Probe pair 1 is blue and binds to exon 7. Probe pair 2 is red and binds to exons 5 and 7. In splice variants lacking exon 6, probe pair 2 binds close enough for FRET to occur. The black arrows are the amplification primers

exon 7. Therefore, fluorescence resonance energy transfer (FRET) occurs between the probes only when exon 6 is missing.

The mastermix shown in Table 2 was used for the amplification and hybridization probe-based detection of Ikaros isoforms.

**LightCycler PCR**

18 μl of the mastermix and 2 μl of the corresponding template solution were added to the LightCycler glass capillary. In each run, a negative control was prepared by replacing the template with water.

The amplification protocol is shown in Table 3. Before amplification, an initial 10-min denaturation at 95°C was performed.

The melting curve procedure in Table 4 was used. Melting curves are not shown.

**Table 2.** Mastermix

|  | Volume [µl] | [Final] |
|---|---|---|
| LightCycler-FastStart DNA master Hybridization Probes | 2 | 1X |
| MgCl$_2$ stock solution | 1,6 | 3 mM |
| Primers (10 µM each) | 1+1 | 0.5 µM |
| Probes (3.3 µM each) | 1+1+1+1 | 0.2 µM |
| H$_2$O (PCR grade) | 8,4 |  |
| Total mastermix volume per reaction | 18 |  |

**Table 3.** PCR protocol

| Parameter | Value | | |
|---|---|---|---|
| Cycles | 40 | | |
|  | Segment 1 | Segment 2 | Segment 3 |
| Target temperature [°C] | 95 | 52 | 72 |
| Incubation time [s] | 15 | 15 | 16 |
| Acquisition mode | None | Single | None |

## Results

We describe here a method to measure the relative expression of chicken Ikaros isoforms. Probe pair 1 was designed to detect the total amount of all Ikaros isoforms. It binds to exon 7, which is present in all splice variants. Probe pair 2 is designed with the donor probe at the 3' end of exon 5 and the acceptor probe at the 5' end of exon 7. FRET occurs only when the probes bind to an isoform lacking exon 6. Hence, this probe pair specifically measures the amount of isoforms lacking exon 6 (Figure 2).

**Specificity Test**   The probe pairs' specificity was analyzed with plasmid DNA containing the coding sequence of Ikaros isoforms Ik-1 and Ik-1A (Figure 3). 1x10–3 dilutions and different mixes of Ik-1 and Ik-1A were used as a template. Probe pair 1 measured all isoforms, and fluorescence in channel 705 nm could be measured in all reactions except the negative control. The probe pair 2 showed no fluorescence in reactions containing only Ik-1. Fluorescence at 640 nm was detected only in reactions containing Ik-1A. The interference of Ik-1 in the same capillary with Ik-1A was characterized with various isoform mixtures. The results showed that the detection limit of isoforms lacking exon 6 is 1% of the total isoform concentration.

**Analysis of Ikaros Isoforms in Chicken Embryos and in Cell Lines**   As an example for the use of this method, we have studied the Ikaros isoform expression from total embryos on embryonic day 2 and 5, and from various B and T cell lines. A standard curve was created with serial dilutions of chicken DT40 B cell line cDNA. The amount of all Ikaros isoforms and those lacking exon 6 were measured in the same reaction tube (Figure 4). The second derivative maximum

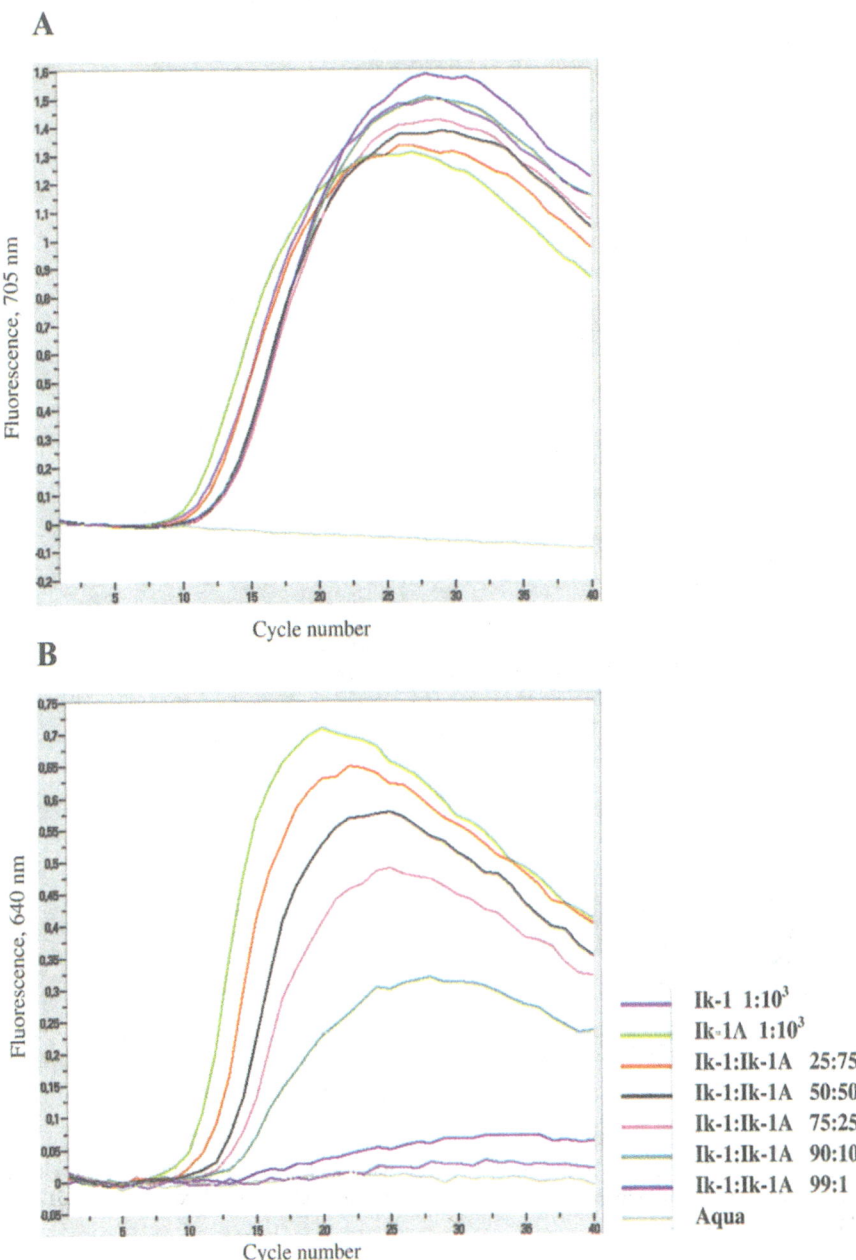

**Fig. 3.** Probe specificity. Plasmids containing the coding sequence of Ikaros isoforms Ik-1 and Ik-1A and various mixes of them were used as a template. (**a**) Fluorescence channel F3, 705 nm, measures the FRET from probe pair 1 of all Ikaros splice variants. (**b**) Fluorescence channel F2, 640 nm, measures the FRET from probe pair 2 of isoforms lacking exon 6

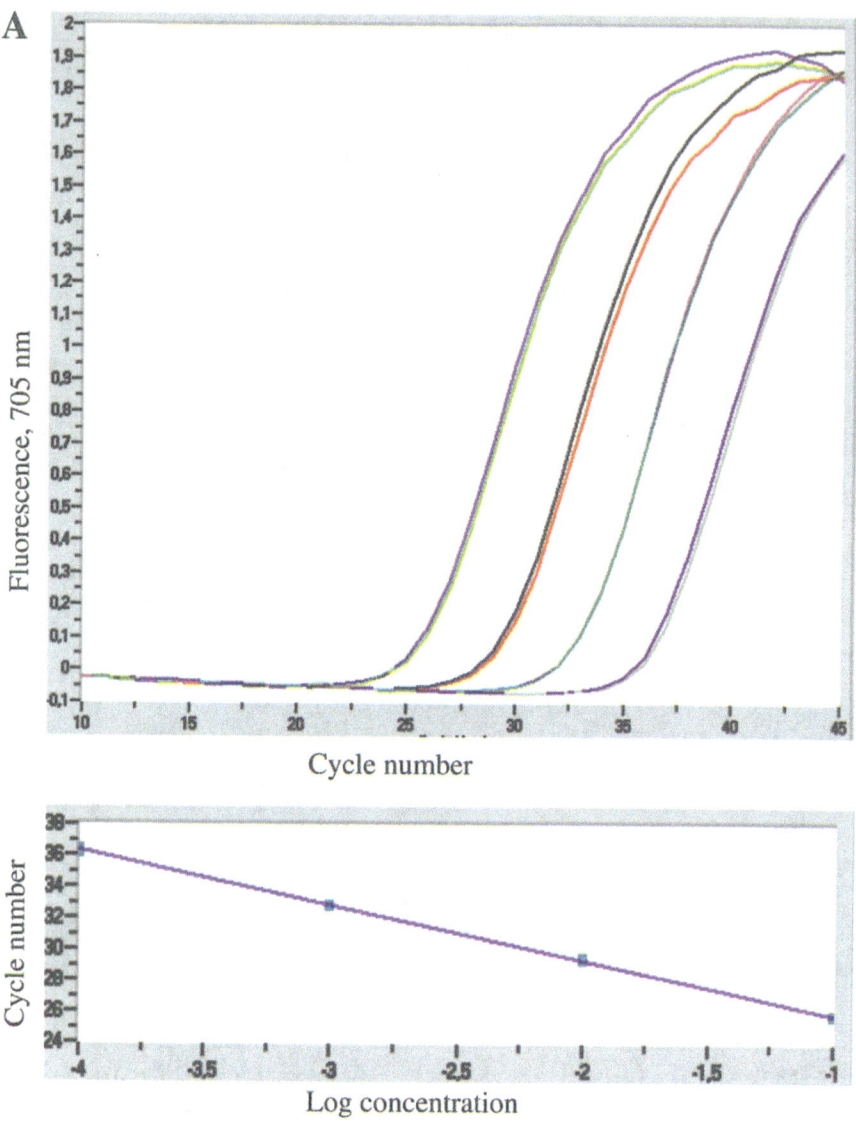

**Fig. 4.** Amplification curves and standard curves made with serial dilutions of DT40 cell line cDNA. **(a)** The total amount of all Ikaros isoforms. The slope of the standard curve is –3.512 and the PCR efficiency 1.93

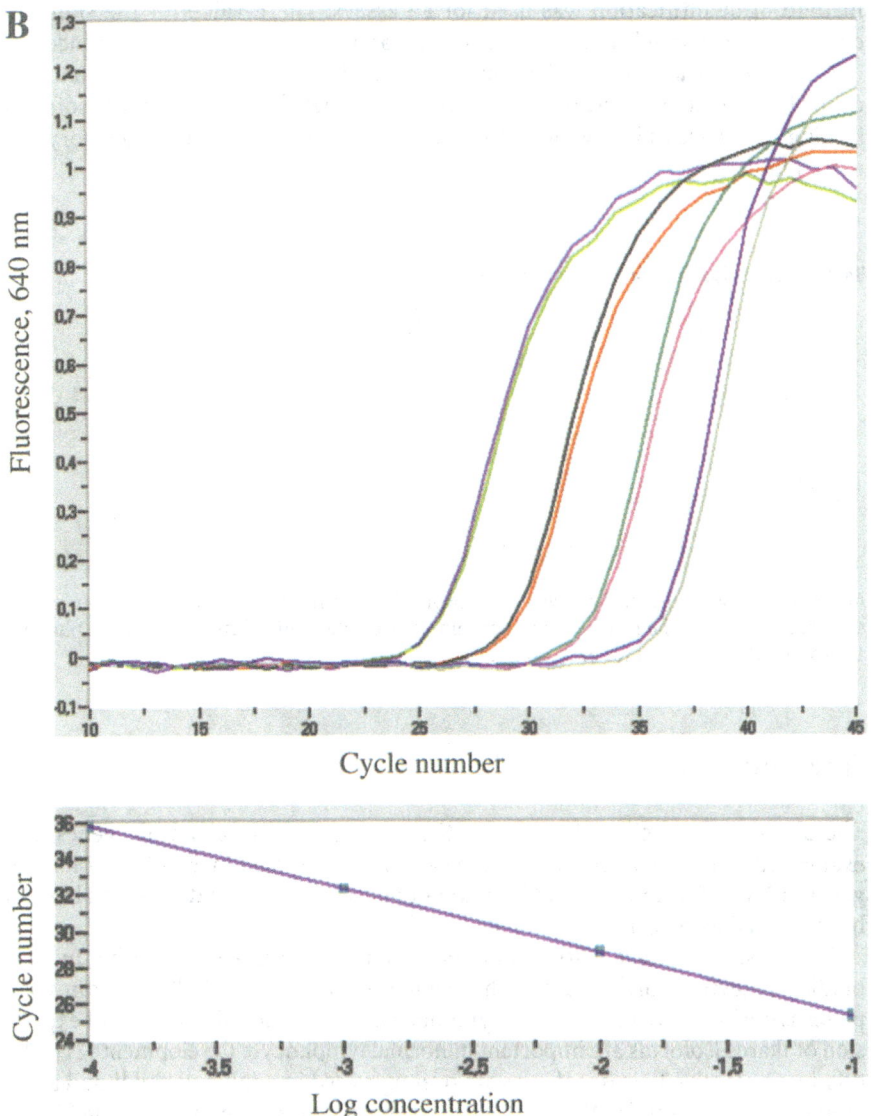

**Fig. 4.** Amplification curves and standard curves made with serial dilutions of DT40 cell line cDNA. **(b)** Isoforms lacking exon 6. The slope of the standard curve is −3.508 and the PCR efficiency 1.93

method of quantification was used for all assays. The E-value for the standard curve in both channels was 1.93 and for the samples 1.88–1.98. The difference was due to different purities of the templates. β-actin expression was detected as a control as previously described [12]. We found that the expression of isoforms lacking exon 6 was slightly increased at embryonic day 5 compared to embryonic day 2 (Table 4).

**Table 4.** Ikaros isoforms in chicken embryos

| Cells | Ik-1A and Ik-2A / All Ikaros isoforms |
|---|---|
| Embryonic day 2 | 1.005 ±0.059 |
| Embryonic day 5 | 1.142 ±0.017 |
| DT40 | 1.000 ±0.038 |
| RP-13 | 0.974 ±0.047 |
| Cu-17 | 0.987 ±0.017 |
| Cu-32 | 0.965 ±0.037 |
| Cu-36 | 1.051 ±0.085 |

Mean±S.D. of four individual analysis are shown. cDNA from DT40 cell line was used to create standard curves and the relative values for samples were determined from the curve. Dilutions 1: 10 and 1:100 were used for each sample

## Comments

The alternative splicing of nuclear RNA is common in eukaryotic genes. For example, at least one alternative splice variant is expressed in 40–60 % of human genes [17, 18, 19]. However, the function of alternatively spliced isoforms has not been studied extensively.

Ikaros isoforms have distinct binding affinities, differences in specific *in vitro* binding sequences, differences in the cellular localization, and altered expression patterns during the development of lymphocytes. [1, 20]. Certain patterns of expression of Ikaros isoforms are important in normal lymphocyte development [21]. The dominant negative Ikaros isoforms are overexpressed in adult and childhood acute lymphoblastic leukemia [10, 11, 22, 23], and chronic myelogenous leukemia [24].

The binding sites of the interaction proteins in Ikaros molecule have been characterised (9) but the function of the highly conserved area in exon 6 still remains unknown. We found a minor alteration of Ikaros isoform expression during embryonic development. The relative expression of splice variants lacking exon 6 was increased at a later stage of development. This is parallel to the results of Klug (20) who showed Ikaros isoform alterations in hematopoietic stem cells. With the method described here, we will be able to analyse the alterations of the expression of isoforms lacking exon 6 in various developmental stages, in different cell types and in malignant cells. We described earlier a hybridization probe-based method for measuring Aiolos splice variants [12]. For the Ikaros isoforms, we have made some improvements to the method. Dual color analysis, where both probe pairs are in the same reaction tube, increases the reliability of results by decreasing the effects of pipeting errors and variations in reaction conditions. This method provides an effective tool to characterize the expression patterns of splice variants of almost any isoform-forming gene.

*Acknowledgements*
We want to thank Pia Nieminen and Kimmo Koskela for kindly providing cDNAs for the analysis.

# References

1. Georgopoulos, K., Winandy, S. and Avitahl, N. (1997) The role of the Ikaros gene in lympho-cyte development and homeostasis. Annu Rev Immunol. 15:155
2. Kelley, C.M., Ikeda, T., Koipally, J., Avitahl, N., Wu, L., Georgopoulos, K. and Morgan, B. (1998) Helios, a novel dimerisation partner of Ikaros expressed in the earliest hematopoietic pro-genitors. Curr Biol 8:508
3. Morgan, B., Sun, L., Avitahl, N., Andrikopoulos, K., Ikeda, T., Gonzales, E., Wu, P., Neben, S. and Georgopoulos, K. (1997) Aiolos, a lymphoid restricted transcription factor that interacts with Ikaros to regulate lymphocyte differentiation. EMBO J 8:2004
4. Koipally, J., Renold, A., Kim, J. and Georgopoulos, K. (1999) Repression by Ikaros and Aiolos is mediated through histone deacetylase complexes. EMBO J 18:3090
5. Liippo, J., Mansikka, A. and Lassila, O. (1999) The evolutionary conserved avian *Aiolos* gene encodes alternative isoforms. Eur J Immunol 29:2651
6. Liippo, J., Nera, K-P., Kohonen, P., Lampisuo, M., Koskela, K., Nieminen, P. and Lassila, O. (2000) The Ikaros family and the development of early intraembryonic hematopoietic stem cells. Curr Top Microbiol Immunol 251:51
7. Liippo, J., Nera, K-P., Veistinen, E., Lähdesmäki, A., Postila, V., Kimby, E., Riikonen, L., Ham-marstrom, L., Pelkonen, J. and Lassila, O. (2001) Both normal and leukemic B lymphocytes express multiple isoforms of the human *Aiolos* gene. Eur J Immunol 31:3469
8. Nakase, K., Ishimaru, F., Fujii, K., Tabayashi, T., Kozuka, T., Sezaki, N., Matsuo, Y. and Harada, M. (2001) Overexpression of novel short isoforms of Helios in a patient with T-cell acute lymphoblastic leukaemia. Exp Hematol 30:313
9. Georgopoulos, K. (2002) Heamatopoietic cell-fate decisions, chromatin regulation and Ikaros. Nature Rev Immunol 2:162
10. Sun, L., Heerema, N., Crotty, L., Wu, X., Navara, C., Vassilev, A., Sensel, M., Reaman, G.H. and Uckun, F.M. (1999a) Expression of dominant-negative and mutant isoforms of the antileukemic transcription factor Ikaros in infant acute lymphoblastic leukemia. Proc Natl Acad Sci USA 96:680
11. Sun, L., Crotty, M-L., Sensel, M., Sather, H., Navara, C., Nachman, J., Steinherz, P.G., Gayon, P.S., Seibel, N., Mao, C., Vassilev, A., Reaman, G.H. and Uckun, F.M. (1999b) Expression of domi-nant negative Ikaros isoforms in T-cell acute lymphoblastic leukemia. Clin Cancer Res 5:2112
12. Veistinen, E., Liippo, J. and Lassila, O. (2002) Quantification of human Aiolos splice variants by real-time PCR. J Immunol Methods 271:113
13. Buerstedde , J. M. and Takeda, S. (1991) Increased ratio of targeted to random integration after transfection of chicken B cell lines. Cell 67:179
14. Baba, T. W., Giroir, B.P. and Humphries, E. H. (1985) Cell lines derived from avian lymphomas exhibit two distinct phenotypes. Virology 144:139
15. Nazerian, K., Witter, R. L., Crittenden, L. B., Noori-Dalloii, M. R. and Kung, H. J. (1982) An IgM-producing B lymphoblastoid cell line established from lymphomas induced by a non-defective reticuloendotheliosis virus. J Gen Virol 58:351

16. Shat, K.A., (1991) Importance of cell-mediated immunity in Marek's disease and other viral tumor diseases. Poult Sci 70:1165

17. Modrek, B., Resch, A., Grasso, C. and Lee, C. (2001) Genome-wide detection of alternative splicing in expressed sequences of human genes. Nucleic Acids Res 29:2850

18. Modrek, B. and Lee, C. (2002) A genomic view of alternative splicing. Nature Genet 30:13

19. Mironov, A.A., Fickett, J.W. and Gelfand, M.S. (1999) Frequent alternative splicing of human genes. Genome Res 9:1288

20. Klug, C.A., Morrison, S.J., Masek, M., Hahm, K., Smale, S.T. and Weissman, I.L. (1998) Hematopoietic stem cells and lymphoid progenitors express different Ikaros isoforms, and Ikaros is localised to heterochromatin in immature lymphocytes. Proc Natl Acad Sci USA 95:657

21. Tonnelle, C., Bardin, F., Maroc, C., Imbert, A-M., Campa, F., Dalloul, A., Schmitt, C. and Chabannon, C. (2001) Forced expression of the Ikaros 6 isoform in human placental blood CD34+ cells impairs their ability to differentiate toward the B-lymphoid lineage. Blood 98:2673

22. Nakase, K., Ishimaru, F., Avitahl, N., Dansako, H., Matsuo, K., Fujii, K., Sezaki, N., Nakayama, H., Yano, T., Fukuda, S., Imajoh, K., Takeuchi, M., Miyata, A., Hara, M., Yasukawa, M., Takahashi, I., Taguchi, H., Matsue, K., Nakao, S., Niho, Y., Takenaka, K., Shinagawa, K., Ikeda, K., Niiya, K. and Harada, M. (2000) Dominant negative isoform of the Ikaros gene in patients with adult B-cell acute lymphoblastic leukemia. Cancer Res 60:4062

23. Olivero, S., Maroc, C., Beillard, E., Gabert, J., Nietfeld, W., Chabannon, C. and Tonnelle, C. (2000) Detection of different Ikaros isoforms in human leukaemias using real-time quantitative polymerase chain reaction. Br J Haematol 110:826

24. Nakayama, H., Ishimaru, F., Avitahl, N., Sezaki,N., Fujii, K., Nakase, K., Ninomiya, Y., Harashima, A., Minowada, J., Tsuchiyama, J., Imajoh, K., Tsubota, T., Fukuda, S., Sezaki, T., Kojima, K., Hara, M., Takimoto, H., Yorimitsu, S., Takahashi, I., Miyata, A., Taniguchi, S., Tokunaga, Y., Gondo, H., Niho, Y., Nakao, S., Kyo, T., Dohy, H., Kamada, N. and Harada, M. (1999) Decreases in Ikaros activity correlate with blast crisis in patients with chronic myelogenous leukemia. Cancer Res 59:3931

25. Hahm, K., Ernst, P., Lo, K., Kim, G., Turck, C. and Smale, S. (1994) The lymphoid transcription factor LyF-1 is encoded by specific, alternatively spliced mRNAs derived from the Ikaros gene. Mol Cell Biol 14:7111

26. Molnár, Á. and Georgopoulos, K. (1994) The Ikaros gene encodes a family of functionally diverse zinc finger DNA-binding proteins. Mol Cell Biol 14:8292

27. Hansen, J.D., Strassburger, P. and Du Pasquier, L. (1997) Conservation of master hematopoietic switch gene during vertebrate evolution: isolation and characterization of Ikaros from teleost and amphibian species. Eur J Immunol 27:3049

28. Georgopoulos, K., Moore, D. and Defler, B. (1992) Ikaros, an early lymphoid-specific transcription factor and a putative mediator for T cell commitment. Science 258:808

# Methods to Quantify Cytokine Gene Expression by Real-Time PCR

PATRICK STORDEUR*

## Introduction

Immune response implies the participation of different cells including antigen presenting cells (such as dendritic cells, monocytes, and macrophages), CD4 and CD8 T lymphocytes, B lymphocytes, and granulocytes. The cooperation between these cells occurs by direct cell-cell contact and by the secretion of proteic factors called cytokines. These cytokines are produced rapidly and transiently, resulting in a fine regulation of the immune response. For example, tumour necrosis factor-alpha (TNF-$\alpha$) is secreted by activated monocytes/macrophages and favours inflammation. In contrast, interleukin-10 (IL-10) is an inhibitory cytokine that limits inflammatory reactions. Interestingly, TNF-$\alpha$ is able to induce the production of IL-10, which in turn inhibits TNF-$\alpha$ secretion by the same cells. This "cytokine network" plays a crucial regulatory role in host defence, making its study particularly useful to understand molecular mechanisms of immune-related diseases [1–2].

The rapid and transient production of cytokines is due to the fine regulation of their gene transcription and/or mRNA stability. An easy method to analyze cytokine production is to quantify their corresponding mRNA by RT-PCR [3]. Because of the accurate quantification provided by real-time PCR, we adapted this technology to cytokine mRNA quantification [4], first from cultured cells, and secondly from whole blood [5]. Whole blood presents several advantages compared to PCR performed from purified cells: it is more representative of the *in vivo* state and takes less time.

## Materials

GeneAmp® PCR System 2400 thermal cycler (Applied Biosystems)      Equipment
Ultrospec® 2000 spectrophotometer (Amersham Pharmacia Biotech)
MagNA Pure LC instrument (Roche Diagnostics)
LightCycler® instrument (Roche Diagnostics)

* Patrick Stordeur, Hôpital Erasme, Laboratoire d'Immunologie. Université Libre de Bruxelles, 808 route de Lennik, 1070 Bruxelles. Belgium.
E-mail: Patrick.Stordeur@ulb.ac.be

**Reagents**
Wizard® SV Gel Clean-up System (Promega Benelux, Leiden, The Netherlands)
MagNA Pure LC mRNA Isolation Kit I (Roche Diagnostics)
LightCycler® – RNA Master Hybridization Probes Kit (Roche Diagnostics)
Oligonucleotides (Biosource Europe, Nivelles, Belgium, and Proligo, Paris, France)
PAXgene™ tubes (Qiagen, Westburg, the Netherlands)

## Procedure

**Sample Preparation**

Purified cells (100,000 to 400,000 peripheral blood mononuclear cells [PBMC], or 50,000 to 100,000 dendritic cells [DC], or 100,000 T cells) suspended in 200 µl of culture medium were lysed and stabilized by adding 500 µl of the PAXgene tube reagent. After vortexing for a few seconds, the lysate is stable for at least five days at room temperature. Whole blood samples were lysed and stabilized using the same method.

**mRNA Extraction**

The lysate from purified cells or whole blood was briefly mixed before transferring a 300-µl aliquot to a 1.5-ml eppendorf tube for centrifugation at 12,000 to 16,000 g for 5 minutes. The supernatant was discarded and the nucleic acid pellet thoroughly dissolved by vortexing in 300 µl of the lysis buffer contained in the MagNA Pure LC mRNA Isolation Kit I. mRNA was then extracted from 300 µl of this solution using the MagNA Pure kit and instrument ("mRNA I cells" Roche's protocol, final elution volume 100 µl).

**One-Step RT-PCR**

Reverse transcription and PCR were performed in one step using the LightCycler RNA Master Hybridization Probes Kit. Five µl of the mRNA preparations were directly added to the PCR reaction without mRNA concentration correction (the latter was too low to be measurable by spectrophotometry).

**LightCycler Master Mix Table**

One-step RT-PCR reaction for each 20-µl reaction:

|  | Volume [µl] | [Final] |
| --- | --- | --- |
| LC-RNA Master Hybridization probes (2.7x) | 7.5 | 1x |
| Mn(OAc)$_2$ (50mM) | 1.3 | 3.25 mM |
| Human β-actin forward primer (6 µM) | 1.00 | 0.3 µM |
| Human β-actin reverse primer (6 µM) | 1.00 | 0.3 µM |
| Human IL-10 forward primer (6 µM) | 2.00 | 0.6 µM |
| Human IL-10 reverse primer (6 µM) | 3.00 | 0.9 µM |
| Human TNF-α forward primer (6 µM) | 3.00 | 0.9 µM |
| Human TNF-α reverse primer (6 µM) | 3.00 | 0.9 µM |
| Hydrolysis probe (4 µM) | 1.0 | 0.2 µM |
| H$_2$O | up to 20 µl |  |

5 μl of purified mRNA preparation or external standard dilution was added to each capillary. The optimal concentration of forward and reverse primers depends on the cDNA target. For TNF-α, the actual total volume is 20.8 μl without water.

The LightCycler protocol was as follows:
- Reverse transcription for 20 min at 61°C
- Denaturation for 30 s at 95°C
- Amplification

| Parameter | Value | |
|---|---|---|
| Cycles | 45 | |
| Type | Quantification | |
| | Segment 1 | Segment 2 |
| Target temperature [°C] | 95 | 60 |
| Incubation time [s] | 0 | 20 |
| Temperature transition rate [°C/s] | 20 | 20 |
| Acquisition | None | Single |
| Gains | | F1=8/F2=2/F3=4 |

**Amplification Parameter Table**

Fluorimeter gains were adjusted using the real-time fluorimeter function of the LightCycler following manufacturer's instructions. We found optimal fluorescent signals for most of the PCRs with the following gain values: F1 = 8, F2 = 2, F3 = 4.

**Fluorimeter Gains Adjustment**

Primers and hydrolysis (TaqMan) probes were designed with Primer 3 software [6]. The default parameters of the program were applied except for the following: product size 110 to 150 bp, primer size 20 to 27 bp, primer Tm 58 to 62°C with a max Tm difference of 2.0°C, product Tm 0 to 85°C, max self and 3' self complementarily = 6.00, max poly-X = 3, primer and Hyb Oligo penalty (penalty weights for primer pairs) = 2.0, Hyb Oligo Tm 68 to 72°C. In addition, the oligonucleotides were selected according to the following criteria (listed in order of importance): (1) intron spanning if possible, (2) no G at the probe 5' end, (3) no more than two Gs or Cs within the five 3'nucleotides for primers, and (4) more Cs then Gs in the probe.

**Oligonucleotide Design**

A standard curve was constructed for each PCR run from serial dilutions of a DNA of known concentration. This DNA was either a purified PCR product (for β-actin and TNF-α), or a plasmid obtained from the American Type Culture Collection (for IL-10). Sample-to-sample variations within a PCR run were corrected against an internal standard (the housekeeping gene β-actin).

**Generation of External Standards**

**Table 1.** Oligonucleotides

| Human β-actin (GenBank Accession # X00351) | | | | |
|---|---|---|---|---|
| | Position | Length | GC (%) | $T_m$ (°C) |
| **External standard primers** | | | | |
| CCCTGGAGAAGAGCTACGA | 745 | 19 | 58 | 58.1 |
| TAAAGCCATGCCAATCTCAT | 1253 | 20 | 40 | 58.2 |
| Product | 745–1253 | 509 | | |
| **Real-time PCR primers** | | | | |
| GGATGCAGAAGGAGATCACTG | 976 | 21 | 52 | 59.8 |
| CGATCCACACGGAGTACTTG | 1065 | 20 | 55 | 59.2 |
| Product | 976–1065 | 90 | | |
| **Hydrolysis probe** | | | | |
| 6F-CCCTGGCACCCAGCACAATG-TAMRA | 997 | 20 | 65 | 69.3 |
| Human IL-10 (GenBank Accession # M57627) | | | | |
| **External standard primers** | | | | |
| TGCCTAACATGCTTCGAGATCT | 140 | 22 | 45 | 61.3 |
| TATAGAGTCGCCACCCTGATGT | 590 | 22 | 50 | 60.9 |
| Product | 140–590 | 451 | | |
| **Real-time PCR primers** | | | | |
| AGCAGGTGAAGAATGCCTTTAA | 449 | 22 | 41 | 60.3 |
| TTCATTGTCATGTAGGCTTCTATG | 554 | 24 | 38 | 58.4 |
| Product | 449–554 | 106 | | |
| **Hydrolysis probe** | | | | |
| 6F-TCCAAGAGAAAGGCATCTACAAAGCCA-TAMRA | 476 | 27 | 44 | 68.7 |
| Human TNF-α (GenBank Accession # M10988) | | | | |
| **External standard primers** | | | | |
| ACCATGAGCACTGAAAGCAT | 83 | 20 | 45 | 58.3 |
| AGATGAGGTACAGGCCCTCT | 488 | 20 | 55 | 57.8 |
| Product | 83–488 | 406 | | |
| **Real-time PCR primers** | | | | |
| CCCAGGGACCTCTCTCTAATC | 275 | 21 | 57 | 59.2 |
| ATGGGCTACAGGCTTGTCACT | 358 | 21 | 52 | 61.0 |
| Product | 275–358 | 84 | | |
| **Hydrolysis probe** | | | | |
| 6F-TGGCCCAGGCAGTCAGATCATC-TAMRA | 303 | 22 | 59 | 69.0 |

# Results

Real-time PCR was developed for each cytokine mRNA by designing two primers and a hydrolysis probe using Primer 3 software (6). Then, two additional primers that encompass the first two primers were designed. These outer primers were used to generate, by conventional PCR, a PCR product purified by the "Wizard SV Gel Clean-up System" following manufacturer's instructions and intended to be used as an external standard (Figure 1). Alternatively, plasmids were used for some cytokine standards. The copy number of the standards was calculated from the DNA concentration measured by spectrophotometry. Using the standard at $10^5$ copies, primer titration was performed at 300, 600 or 900 nM (Figure 2). The expected size of the real-time PCR product was checked by agarose gel electrophoresis.

Because the mRNA obtained after purification with the MagNA Pure was too low to be accurately quantified by spectrophotometry, five μl of this purified mRNA was always used. Therefore, preliminary experiments were needed to determine the optimal amount of starting biological material (Figure 3). The optimal number of PBMC for quantification of TNF-α and β-actin ranged from 100,000 to 600,000 cells. The optimal volume of whole blood ranged from 20 to 200 μl. Quantification of β-actin from CD4 T cells revealed an optimal number for these cells of 100,000 to 1,000,000. To increase sensitivity, the elution volume at the end of mRNA extraction can be reduced to 50 or 25 μl.

Although the correlation between the amount of starting material and the calculated mRNA copy number (r > 0.98, Figure 3) could be used for quantification, it is preferable to correct the measured copies by simultaneous measurement of a housekeeping gene such as β-actin. However, the expression of housekeeping genes may vary with different conditions of stimulation. Furthermore, standardization against housekeeping genes needs to consider the efficiency of both the

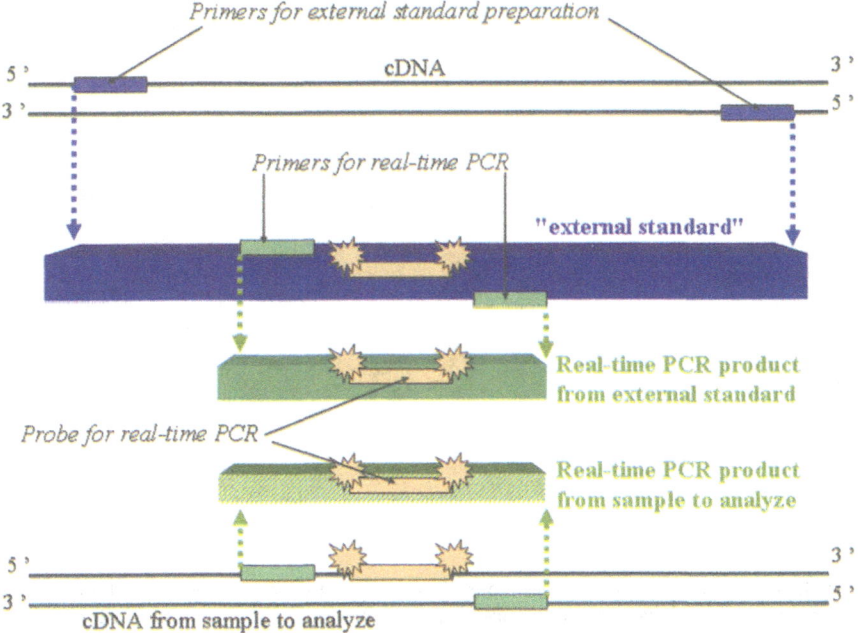

**Fig. 1.** Schematic for real-time PCR and external standard preparation

target mRNA PCR and the housekeeping gene PCR. These techniques are described in detail elsewhere [4].

The protocol described here allows analysis of *in vitro* cytokine mRNA modulation. For example, Figure 4 demonstrates mRNA induction of IL-10 in PBMC by the calcium ionophore A23187 and phorbol myristate acetate. Figure 5 demonstrates TNF-α induction, by lipopolysaccharides in whole blood.

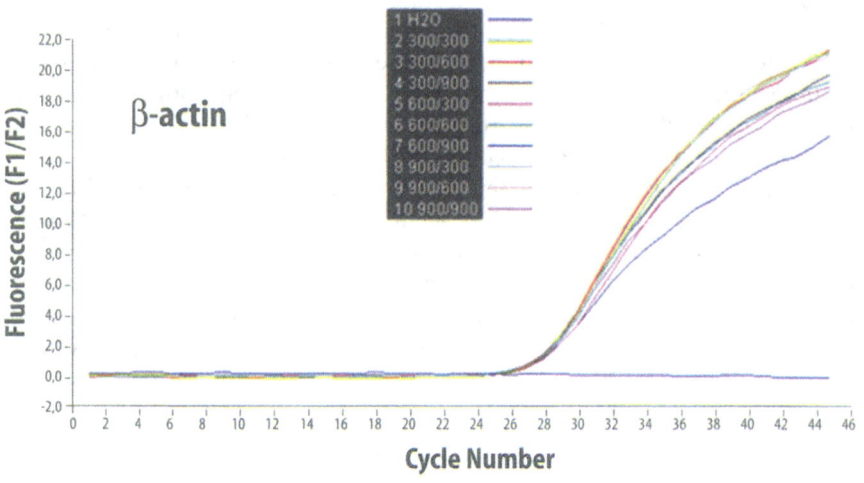

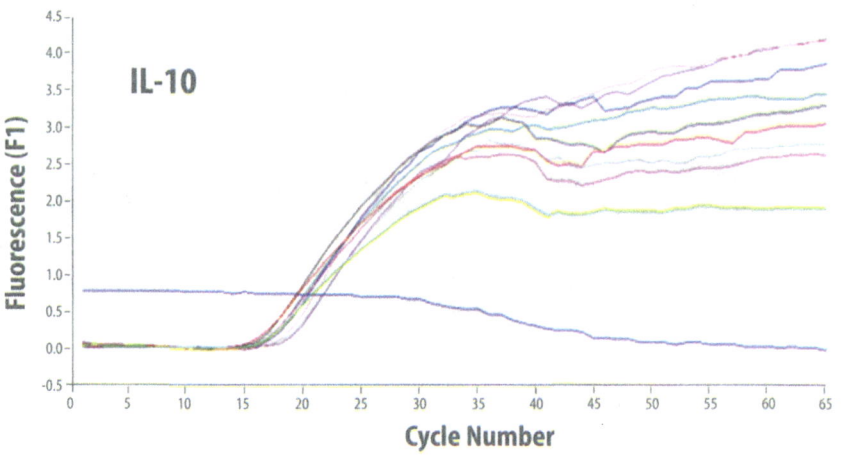

**Fig. 2.** IL-10 and β-actin primer titrations. Three different concentrations of each primer was tested: 300, 600, and 900 nM. Each concentration of forward primer was tested in combination with each concentration of reverse primer in a real-time PCR reaction starting with $10^5$ copies of purified standard. The preferred primer concentrations gave the curve with the highest slope, and the lowest Cp value. Results of TNF-α primer titrations have been previously published [4]

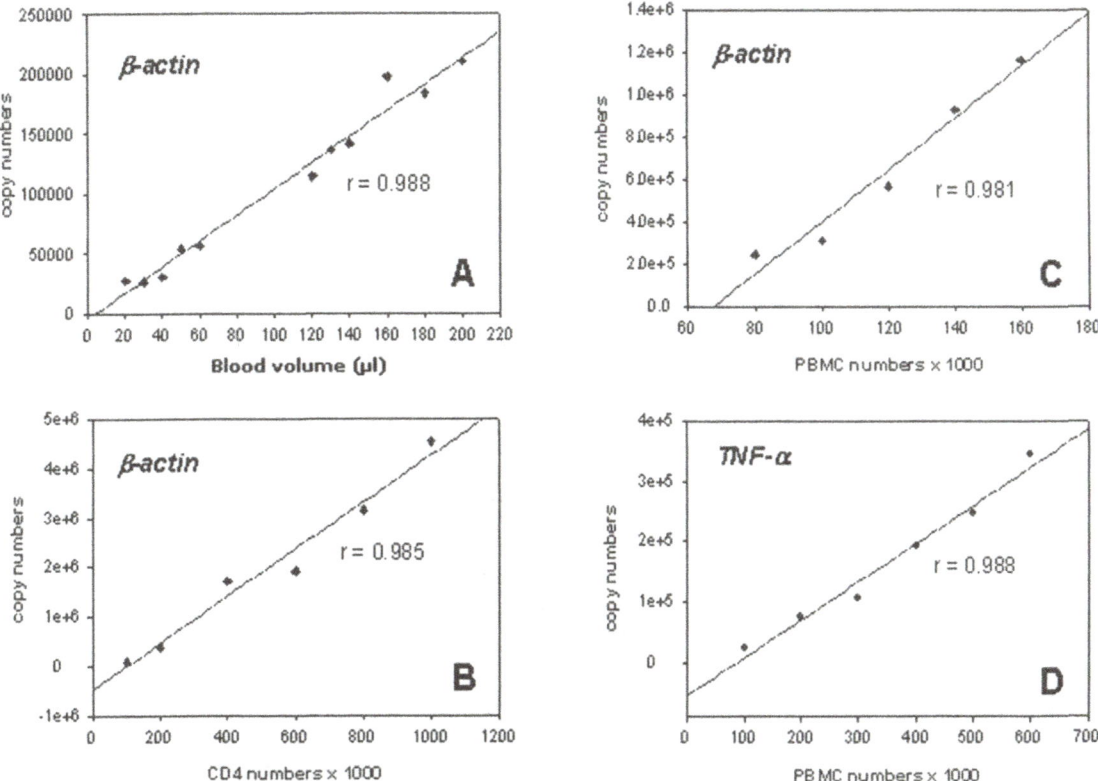

**Fig. 3.** Determination of optimal cell number or blood volume. The Y-axis indicates the copy number and the X-axis the starting cell number or blood volume. The line indicates the linear regression. **A.** Various whole blood volumes were cultured in the presence of 10 ng/ml LPS for six hours. At the end of the culture, RT and real-time PCR for β-actin mRNA was performed. **B, C, D.** Real-time PCR for β-actin mRNA was performed from various numbers of CD4 cells (B) or PBMC (C). PBMCs were also analyzed for TNF-α mRNA (D)

## Comments

A simple strategy to perform and develop quantitative real-time PCR for cytokine mRNA quantification is presented. A single protocol is used for different target mRNAs, the only adjustment being the primer concentration, so that real-time PCR for a new mRNA target may be rapidly developed. The technique should be applicable to any transcript. In our laboratory we have developed real-time PCR for human and mouse transcription factors, membrane receptors, housekeeping genes, chemokines, and chemokine receptors. Using the same protocol for both cultured cells and whole blood, i.e., the use of the reagent contained in the PAX-gene tube, is an advantage of the technique. Cultured cells may be directly lysed

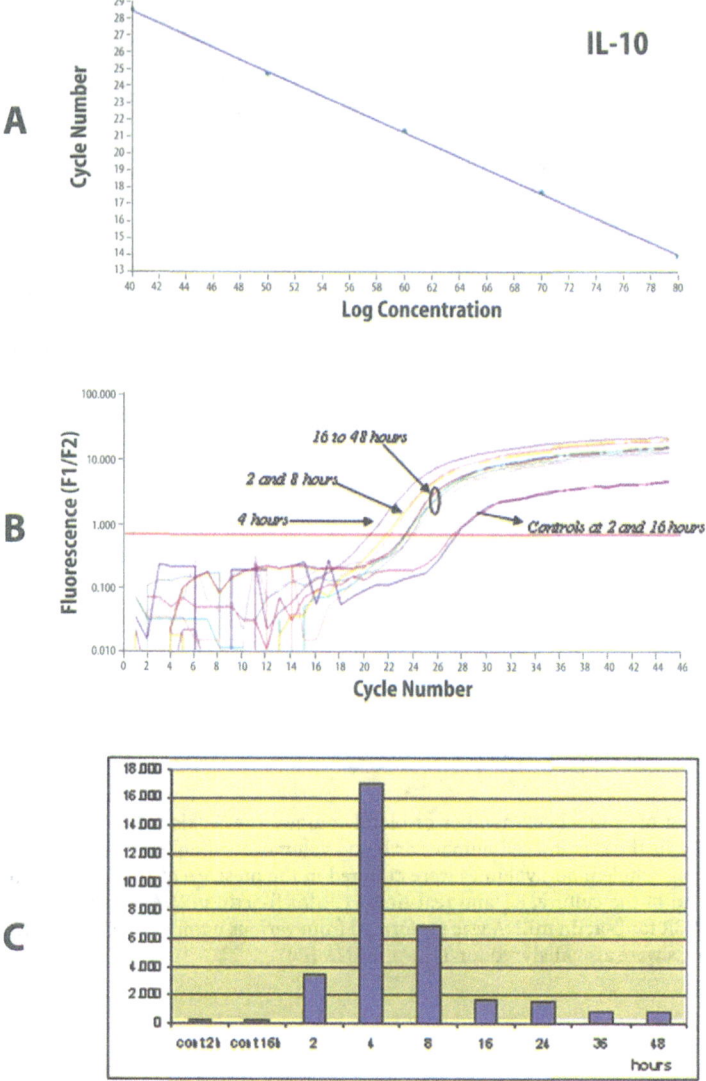

**Fig. 4.** Induction of IL-10 mRNA in PBMC by calcium ionophore A23187 and PMA. 200,000 PBMC were cultured in 200 µl of culture medium, with or without 100 ng/ml calcium ionophore A23187 and 10 ng/ml PMA, for different lengths of time. Cultures were stopped by adding 500 µl of PAXgene tube reagent. Real-time PCRs were performed for β-actin and IL-10 mRNAs. **A.** Standard curve for IL-10 mRNA quantification. Baseline Adjustment: arithmetic; Analysis Method: fit points; Number of Fit Points: 4, Slope: -3.610; Error: 0.0550; r: -1.00, Color Compensation: off. **B.** Fluorescence curves of the samples. "Controls at 2 and 16 hours" are for PBMC cultured in medium alone for 2 and 16 hours. "LightCycler Noise Band Report" is shown. **C.** Results of the experiment. The Y-axis represents IL-10 mRNA copy numbers per million copies of β-actin mRNA, and the X-axis the time of culture. The two first columns are for PBMC cultured in medium alone. A rapid and transient induction of IL-10 mRNA was demonstrated

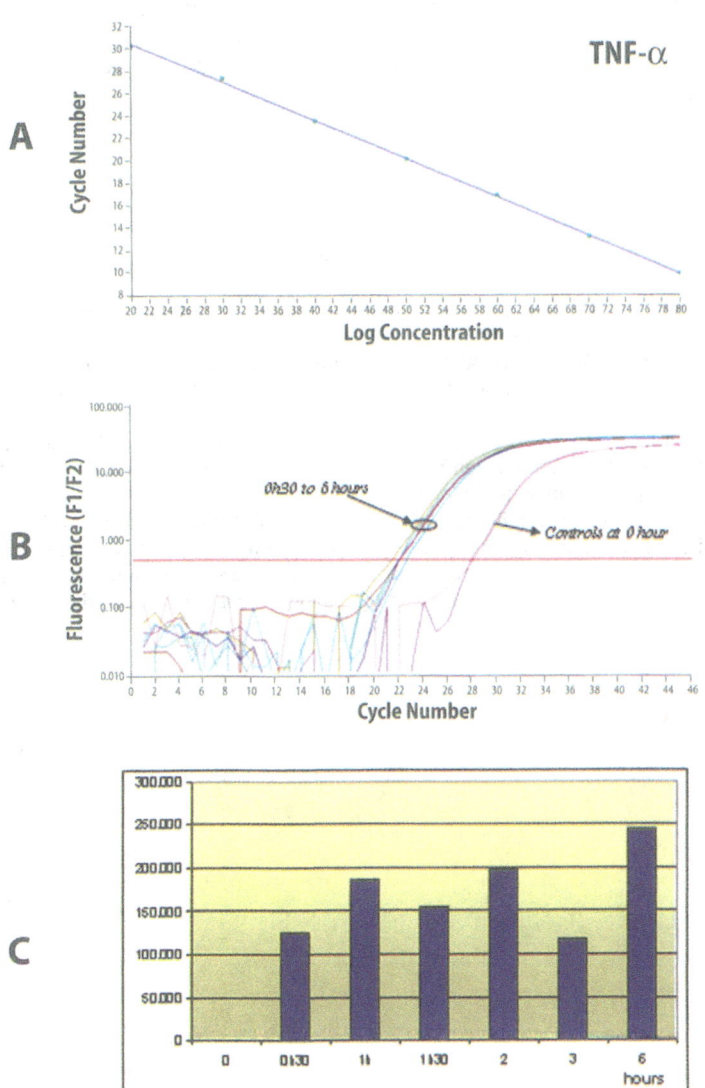

**Fig. 5.** Induction of TNF-α mRNA in whole blood by LPS. 200 µl of blood were cultured in the presence of 10 ng/ml LPS for different lengths of time. Cultures were stopped by adding 500 µl of PAXgene tube reagent. Real-time PCRs were performed for β-actin and TNF-α mRNAs. **A.** Standard curve for TNF-α mRNA quantification. Baseline Adjustment: arithmetic; Analysis Method: fit points; Number of Fit Points: 2, Slope: -3.407; Error: 0.194; r: -1.00, Color Compensation: off. **B.** Fluorescence curves of the samples. "Controls at 0 hour" is for whole blood lysed at time zero, i.e., without LPS activation. "LightCycler Noise Band Report" is shown. **C.** Results of the experiment. The Y-axis represents TNF-α mRNA copy numbers per million copies of β-actin mRNA, and the X-axis the time of culture. A rapid and sustained induction of TNF-α mRNA was demonstrated

in the culture vessel without the need to discard the supernatant, and thus, without cell loss. Finally, whole blood can be collected directly in the PAXgene tube, allowing the accurate quantification of peripheral blood mRNA present at the moment of blood collection [5].

## References

1. Cavaillon JM (2001) Pro- versus anti-inflammatory cytokines: myth or reality. Cell Mol Biol 47:695–702
2. Kourilsky P, Truffa-Bachi P (2001) Cytokine fields and the polarization of the immune response. Trends Immunol 22:502–509
3. Giulietti A, Overbergh L, Valckx D, Decallonne B, Bouillon R, Mathieu C (2001) An overview of real-time quantitative PCR: applications to quantify cytokine gene expression. Methods 25:386–401
4. Stordeur P, Poulin LF, Craciun L, Zhou L, Schandene L, de Lavareille A, Goriely S, Goldman M (2002) Cytokine mRNA quantification by real-time PCR. J Immunol Methods 259:55–64 (Erratum in: (2002) J Immunol Methods 262:229)
5. Stordeur P, Zhou L, Byl B, Brohet F, Burny W, de Groote D, van der Poll T, Goldman M (2003) Immune monitoring in whole blood using real-time PCR. J Immunol Methods 276:69–77
6. Rozen S, Skaletsky H (2000) Primer3 available at http://www.genome.wi.mit.edu/cgi-bin/primer/primer3_www.cgi for general users and for biologist programmers. Methods Mol Biol 132:365

# Applications

**II**

## Oncology

# Profiling Breast Cancer Using Real-Time Quantitative PCR

Scot G. Frank, Philip S. Bernard*

## Introduction

We are at the threshold of a new era in medicine due to advances in technology, informatics and the biological sciences. Complex diseases such as cancer are being deciphered using new tools to analyze genomes, transcriptomes and proteomes. The potential of these discoveries to improve health care is clear, but the best way to implement these findings into the clinical setting remains debatable. Cancer is a heterogeneous disease with respect to cellular make-up, molecular alterations and clinical behavior. The human eye can rapidly discern differences in cell morphology using the light microscope; thus, histochemistry is the main method employed for making a diagnosis of cancer in solid tissues. Unfortunately, anatomic staging strategies with histochemistry alone fail to meet the diversity of molecular alterations that contribute to differences in clinical behavior. Molecular classifications help to stratify cancers into more homogenous groups that better predict an individual's outcome and response to therapy [1].

Immunohistochemistry (IHC) is commonly used clinically for protein expression analysis. The technique is popular because it is inexpensive, performed on formalin-fixed, paraffin-embedded tissue, and allows the marker to be simultaneously associated with histology. Limitations to IHC include being semi-quantitative, low-throughput, and subjective. Moreover, discrepancies in results can arise due to a lack of uniformity in reagents, storage, and criteria used for scoring [2]. In breast cancer, IHC is used to determine protein expression of the hormonal receptors (*ER* and *PgR*) and *HER2/ERBB2* [3]. The presence of hormonal receptors indicates a good prognosis and predicts response to tamoxiten [4], while high expression of *HER2* portends a poor prognosis but patients often show some response to regimens with Herceptin [5]. Due to inconsistencies in IHC testing for *HER2*, evaluation of *HER2* gene (DNA) amplification by fluorescence *in-situ* hybridization (FISH) may be preferred. There is generally poor agreement between FISH and IHC [6], especially for those samples scored weak positive by IHC where only 20% are positive by FISH [7]. Like IHC, FISH can be done on

---

* Philip S. Bernard, University of Utah, Department of Pathology, Rm. 3345 Huntsman Cancer Institute, 2000 Circle of Hope, Salt Lake City, UT 84112–5550
E-mail: phil.bernard@hci.utah.edu

archived tissue and is combined with visualization of the histology. FISH is more quantitative than IHC but labor intensive and expensive [8].

Microarrays are now used in research to compare differences in biological systems on a genomic scale. In cancer research, microarrays coupled with statistical analysis have found gene expression "signatures" that define biological differences between tumors, and predict clinical outcome [1]. Although expression classifications offer strong evidence that a molecular taxonomy for cancer will improve the standard of medical care, there are some impediments to using RNA markers in the clinical lab. For instance, expression profiling using RNA will require significant changes in specimen processing to keep markers from degrading. Many genes that are highly overexpressed, however, are frequently gene amplified and may be scored using DNA. For example, microarray studies using breast tumors and cell lines have shown that 40–60% of genes that are highly overexpressed are DNA amplified [9]. Microarrays are a powerful tool for scanning the genome and initially defining markers of interest but they may not be optimal for clinical testing since they are labor intensive, and difficult to automate and standardize.

Real-time quantitative PCR can accurately determine gene dosages and these systems are often homogenous, which reduces labor time, risk of contamination, and clerical errors. The method is attractive for molecular diagnostics since it is objective, rapid, versatile, cost-effective, and can be performed on small tissue samples. Sequence specific probes, such as hybridization probes, exonuclease probes, and hairpin probes, can increase fluorescence signal detection during amplification and allow multiplexing. However, this increase in complexity over dsDNA dyes (*e.g.*, SYBR Green I) comes at a higher cost, more optimization, and without appreciable differences in sensitivity [10]. Furthermore, any limitations in the specificity of dsDNA dyes can be overcome using post-amplification melting curves or choosing high fluorescence acquisition temperatures during amplification [11]. This chapter illustrates the potential of using real-time quantitative PCR for molecular staging in cancer. We present methods and strategies for the quantitative analysis of RNA and DNA. Examples are provided using two different breast cancer cell lines that differ in gene expression and regions of DNA amplification.

## Materials

**Equipment**

LightCycler instrument and capillaries (Roche Diagnostics, Mannheim, Germany)
LightCycler 3.5 software (Roche Diagnostics)
Agilent 2100 Bioanalyzer (Agilent Technologies, Palo Alto, CA)
QIAquick PCR purification column (Qiagen, Valencia, CA)
Primer Designer 5 software (Scientific and Educational Software, State Line, PA)
TmUtility software (http://www.idahotech.com)

**Reagents**

RNeasy Midi Kit (Qiagen)
DNA Midi Kit (Qiagen)

RNA 6000 Nano LabChip Kit (Agilent Technologies)
Superscript II First-Strand Synthesis System (Invitrogen Life Technologies, Carlsbad, CA)
TE': 10 mM Tris, 0.1 mM EDTA, pH 8.0
Oligonucleotides (Qiagen)
dNTPs (Roche, Indianapolis, ID, USA): dATP, dTTP, dCTP, dGTP, and dUTP
PCR Buffer 3 (30mM $MgCl_2$) (Idaho Technology Inc, SLC, UT)
SYBR Green I (Molecular Probes, Eugene, OR)
Platinum *Taq* DNA Polymerase (Invitrogen Life Technologies, Carlsbad, CA)
Breast cancer cell lines MCF7 (cat # HTB-22) and SKBR3 (cat # HTB-30) (ATCC, Manassas, VA)

## Procedure

Breast cancer cell lines (MCF7 and SKBR3) were prepared for RNA and DNA using the RNeasy Midi Kit and the DNA Midi Kit following manufacturer's instructions. Integrity of each RNA sample was analyzed on an Agilent 2100 Bioanalyzer using the RNA 6000 Nano LabChip Kit before making cDNA. Briefly, two microliters of total RNA (10 ng/μL) were heated to 70 °C and 1 μL was loaded on the column. Degradation was evaluated using the signal of the 18S and 28S ribosomal peaks. First strand cDNA was synthesized from 1 μg total RNA using oligo-dT primers and Superscript II reverse transcriptase. The reaction was held at 42 °C for 50 min, followed by a 15-min step at 70 °C, and a 20-min step with RNase H at 37 °C. The single-stranded cDNA was bound on a QIAquick PCR purification column and eluted in 80 μl of Elution Buffer. The cDNA was diluted in TE' (~ 5 ng/μl), aliquoted, and stored at –80 °C until further use.

Sequences for targets of interest were downloaded from Human Genome Browser (http://genome.ucsc.edu) or NCBI (http://www.ncbi.nlm.nih.gov/mapview). Evidence Viewer (http://www.ncbi.nlm.nih.gov/entrez) was used to identify intron-exon boundaries and polymorphisms. Primers for RNA targets were designed to exclude DNA amplification by placing forward and reverse primers in different exons that span a relatively large intron (*e.g.,* >1 kbp). DNA primers were positioned to avoid polymorphisms. Primer stability, hairpins and primer-dimer interactions were evaluated for all primer sets using Primer Designer 5 software. $T_m$ values were calculated for PCR buffers using TmUtility. All primer sets were designed to have a $T_m \approx 60$ °C, GC content $\approx$ 50–60% and to generate a PCR amplicon between 100–250 bp. Finally, BLAST searches were performed on primer pair sequences using the Human Genome Browser to check for uniqueness, including the presence of amplifiable pseudogenes. Primer sequences and characteristics are presented in Table 1.

All PCR reactions were performed on the LightCycler using the master mix from Table 2. Temperature profiles for amplification and melting curve analysis are provided in Table 3. The dsDNA dye SYBR Green I was used as the fluorescent probe for all reactions. Each capillary contained 18 μl of master mix and 2 μl of the appropriate DNA or RNA template. Although the PCR uses dUTP, uracyl-N-

**Table 1.** Oligonucleotides

| ERCC2 (UniGene reference Hs.99987) GenBank Accession #L47234 | | | | | |
|---|---|---|---|---|---|
| | Position | Length | GC (%) | $T_m$ (°C) | Purity ($A_{260}/A_{280}$) |
| **Primers DNA** | | | | | |
| GGGCGTTTTCAGGAGAGAC | 35623–641 | 19 | 57.9 | 62.4 | 1.62 |
| CCAGTAGGGACAGGATGTT | 35781–763 | 19 | 52.6 | 60.1 | 1.71 |
| **HER2/ERBB2 (UniGene reference Hs.446352) GenBank Accession #M11730** | | | | | |
| **Primers DNA** | | | | | |
| CGTGCCAGTGTGAACCAGAA | 3906–925 | 20 | 55 | 64.9 | 1.59 |
| CTCTTGATGCCAGCAGAAGT | 3996–977 | 20 | 50 | 62.0 | 1.67 |
| **GRB7 (UniGene reference Hs.86859) GenBank Accession #AF274875** | | | | | |
| **Primers DNA** | | | | | |
| CTCTCCTCTGGCTCAGAACT | 1967–986 | 20 | 55 | 62.3 | 1.75 |
| GGGCTGGTACCCTCAAGAC | 2144–126 | 19 | 63.2 | 63.7 | 1.68 |
| **PSMC4 (UniGene reference Hs.211594) GenBank Accession #NM 153001** | | | | | |
| **Primers RNA** | | | | | |
| GGCATGGACATCCAGAAG | 449–466 | 18 | 55.6 | 59.5 | 1.88 |
| CCACGACCCGGATGAAT | 638–622 | 17 | 58.8 | 60.9 | 2.03 |
| **HER2/ERBB2 (UniGene reference Hs. 446352) GenBank Accession #NM 004448** | | | | | |
| **Primers RNA** | | | | | |
| TGAAACCTGACCTCTCCTAC | 1961–980 | 20 | 50.0 | 60.0 | 1.66 |
| CAGAATGCCAACCACCG | 2136–120 | 17 | 58.8 | 60.5 | 1.68 |
| **ER/ESR1 (UniGene reference Hs.1657) GenBank Accession #NM 000125** | | | | | |
| **Primers RNA** | | | | | |
| CATGATCAGGTCCACCTTCT | 1477–496 | 20 | 50.0 | 60.8 | 1.66 |
| AGCAGCATGTCGAAGATCTC | 1646–627 | 20 | 50.0 | 61.8 | 1.83 |

**Table 2.** LightCycler PCR Master Mix

| | Volume [μl] | [Final] |
|---|---|---|
| PCR Buffer (10X) | 2 | 1X |
| dNTPs (10X) | 2 | 0.2 mM dATP, dCTP, dGTP; 0.1 mM dTTP; 0.3 mM dUTP |
| Primers (60 μM) | 0.13 + 0.13 | 0.4 μM each |
| SYBR Green I (1/3,000) | 1.5 | 1/40,000 |
| Platinum Taq (1U/μl) | 0.2 | 0.2U |
| H$_2$0 (PCR grade) | 12 | |
| Total volume | 18 | |

**Table 3.** Temperature Profiles for PCR Amplification and Melting Curves

| | Cycles | Analysis Mode | Target Temperature °C | Incubation Time (m:s) Rate °C/s | Temperature Transition | Acquisition Mode |
|---|---|---|---|---|---|---|
| Denaturation | 1 | None | 94 | 1:00 | 20.00 | None |
| Cycling | 50 | Quantification | 94 | 0:03 | 20.00 | None |
| | | | 58 | 0:06 | 20.00 | None |
| | | | 72 | 0:06 | 2.00 | Single |
| Melt | 1 | Melting Curves | 94 | 0:15 | 20.00 | None |
| | | | 60 | 0:15 | 3.00 | None |
| | | | 97 | 0:00 | 0.10 | Continuous |

glycosylase (UNG) is only used if contamination is detected. For a negative control, PCR grade $H_2O$ replaced the template. Capillaries were sealed, centrifuged and placed into the LightCycler.

Fluorescence is acquired each cycle during real-time PCR amplification. As PCR product accumulates, fluorescence signal increases and rises above background at a quantifiable point referred to as the Cp. The LightCycler software automatically computes a Cp from the PCR amplification and reports this value as a fractional number representing the 2nd derivative maximum (point of maximum acceleration) on the amplification curve (fluorescence *versus* cycle number). The Cp value is very reproducible under constant conditions (*e.g.*, primer set, enzyme, buffer, dye probes) and corresponds to the starting amount of template. All quantitative analysis were done using Cp values. The efficiency of PCR for each target was calculated using a calibration curve constructed from a plot of Cp *versus* log ng DNA [12].

$$E = 10^{-1/\,slope}$$

The control gene for DNA (*ERCC2*) was chosen since it was shown to have no alterations after analysis of CGH data, comprised of 40 breast tumors and 10 breast cell lines [13]. The control gene for RNA (*i.e.*, housekeeper *PSMC4*) was chosen because it showed little variance in expression over 122 breast samples analyzed by microarray [14]. The relative copy number (or ratio) of a given target gene was compared between two samples using the PCR efficiency of this gene raised to the difference between fractional cycle numbers of these two samples.

$$relative\ copy\ number = E^{\,Cp(control)-Cp(test)}$$

Note that using the relative copy number alone assumes there are no differences in the quantity of starting material and there is no difference in PCR efficiency between target genes. Alternatively, a reference sample can be used to compare each experimental sample for the control and test targets. Any difference in the amount of starting material is then normalized by the results of the reference sample, and different efficiencies between the control and test genes are considered in the calculation.

$$change\ in\ relative\ copy\ number = \frac{E^{Cp(test,REF)-Cp(test,EXP)}}{E^{Cp(control,REF)-Cp(control,EXP)}}$$

For detecting changes in tumor DNA copy number, normal diploid breast cell DNA serves as an appropriate reference. When comparing RNA expression of a test gene across many samples, a common reference of pooled RNA may be used.

## Results

A calibration curve was generated for each target and used to calculate PCR efficiency and a relative starting copy number. Figure 1 shows a typical calibration curve using SYBR Green and the DNA control gene *ERCC2*. With a sensitive and specific primer set, targets can be detected from single cells. For instance, the lowest dilution contains approximately 0.01 ng DNA, which is the amount of DNA contained in only 1–2 cells (7 pg DNA/cell). This allows gene copy numbers to be determined from microdissected samples. In fact, microdissection of solid tumor samples is usually necessary to accurately determine gene copy numbers because the presence of DNA from normal diploid cells can interfere with the quantitative assessment. Although small amounts of DNA can be used for real-time PCR, the accuracy and precision of the quantitative results can be improved using higher concentrations of DNA (*e.g.*, 2 ng/μl) [15].

The y-intercept of the calibration curve is a function of the minimum detectable copy number and can be used to calculate the starting amount of target [12]. When a small number of targets and samples is being tested, it is feasible to perform within-run efficiency curves to determine starting concentrations. This method may be impractical when profiling many markers, in which case efficiencies may be assumed or pre-determined from prior runs.

Figure 2 shows DNA profiling for SKBR3 using the control gene *ERCC2* and the test genes *HER2* and *GRB7*. The experimental sample (SKBR3) was normalized to a reference sample (normal breast DNA). Real-time quantitative PCR results for *HER2* and *GRB7* show that the two genes are equally amplified approximately 8-fold over normal. Both of these genes lie within 10 kb of each other on the 17q21 amplicon, which has been shown to be commonly altered in breast tumors using CGH and cDNA microarray [13, 16]. Interestingly, *HER2* and *GRB7* work cooperatively in the receptor tyrosine kinase pathway to promote cell proliferation and differentiation.

The major advantage that DNA markers have for testing in the clinical lab is that they are stable and can be reliably tested from archived tissue. Although RNA markers are more labile than DNA, expression analysis can provide important information about tumor biology and it is useful for diagnostics. Quantitative

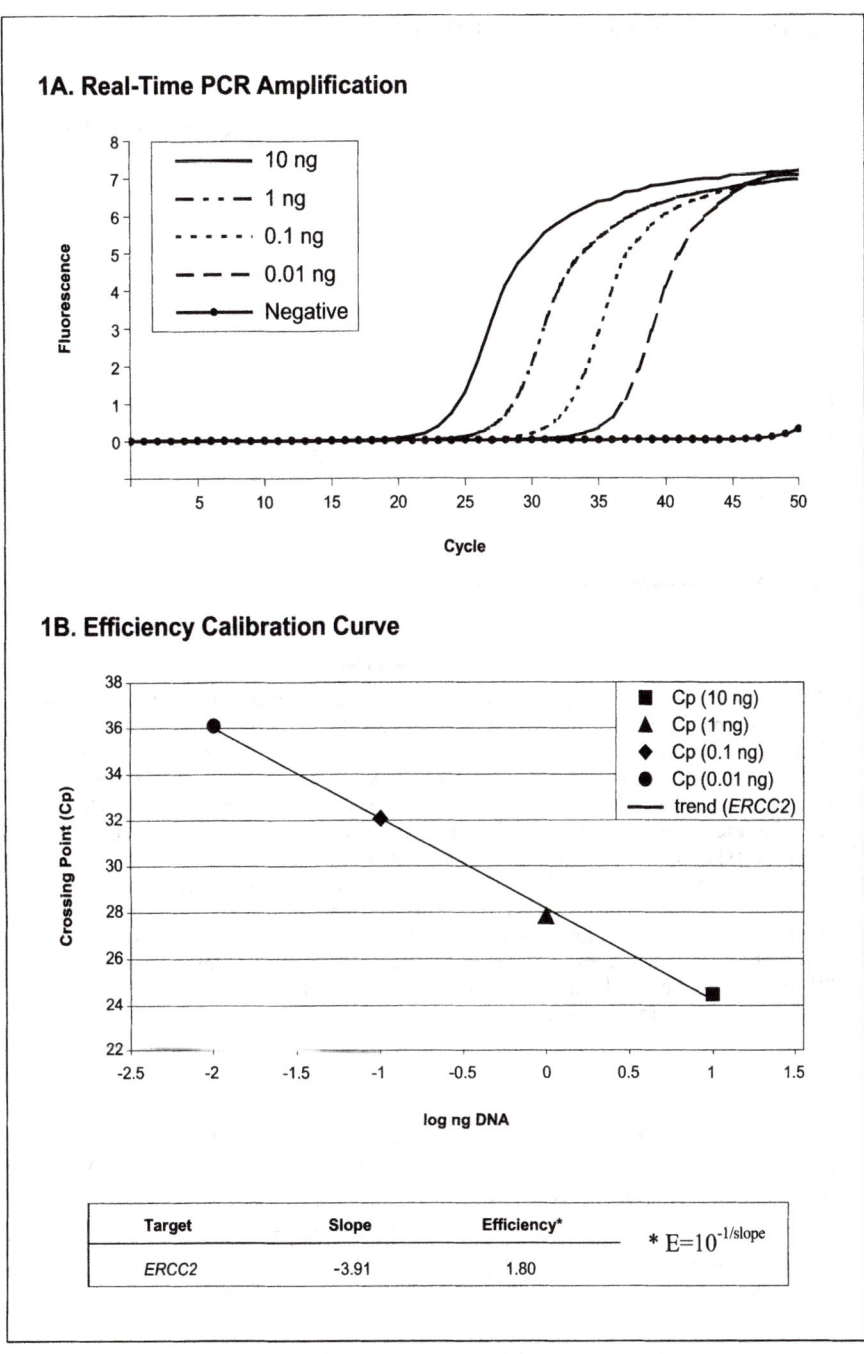

**1A. Real-Time PCR Amplification**

**1B. Efficiency Calibration Curve**

| Target | Slope | Efficiency* | |
|--------|-------|-------------|---|
| ERCC2 | -3.91 | 1.80 | $* E = 10^{-1/slope}$ |

**Fig. 1.** Normal breast DNA was used at decreasing concentrations in a series of 10-fold dilutions. SYBR Green fluorescence was acquired every cycle and then plotted against cycle number to construct an amplification curve at each dilution (**1A**). Linear regression through a plot of the Cps at each dilution against the log of genomic DNA gives the average efficiency of the PCR reaction (**1B**)

## 2. *HER2*, *GRB7* and *ERCC2* in SKBR3

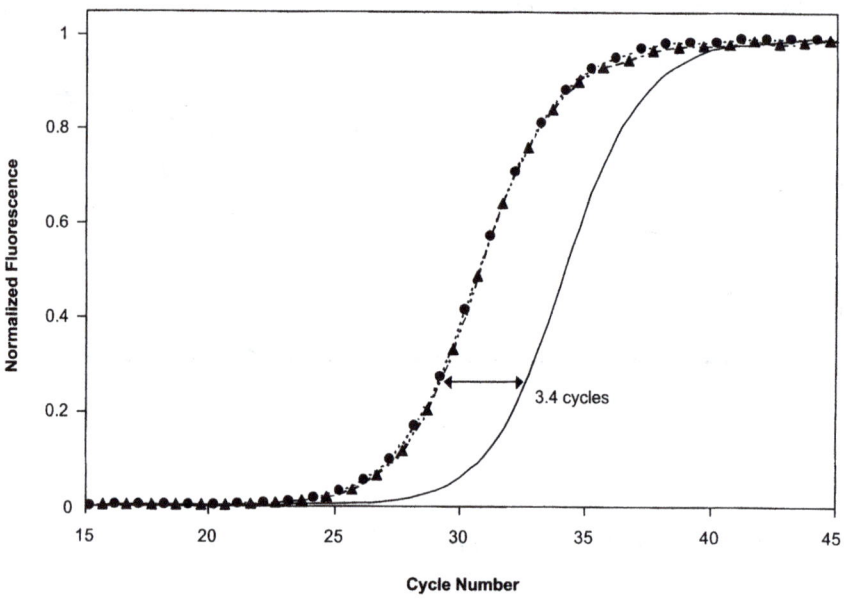

**Fig. 2.** Real-time quantitative PCR for DNA analysis of gene amplification in SKBR3. Genomic DNA from normal breast (reference sample) and SKBR3 (experimental sample) were PCR amplified in the presence of SYBR Green I. The amplification curves for SKBR3 are shown for the control gene *ERCC2* (——) and the test genes *HER2* (- - ● - -) and *GRB7* (- -▲- -). The curves of the test genes (*HER2* and *GRB7*) in SKBR3 are corrected by shifting the cycle number for each fluorescence value in the amplification curve by the observed ΔCp between *ERCC2* and the test gene(s) in the normal breast sample. For instance, if the ΔCp between *ERCC2* and *HER2* was two in the normal sample and four in the tumor sample then the tumor sample would be corrected by two cycles. Note that this is only justified when there is a similar amount of starting material between the samples and similar efficiencies between the targets. Using an average efficiency of 1.85, the cycle shift of 3.4 for *HER2* and *GRB7* shows that both genes are amplified in SKBR3 by approximately 8-fold over normal. This is consistent with other published data for amplification of *HER2* in SKBR3 [17]

analysis of RNA markers for solid tumor profiling in the clinical lab will require implementing pre-analytical methods to assess the integrity of RNA. Figure 3 shows the analysis of three samples with different amounts of RNA degradation. Analyzing the 18S and 28S ribosomal bands provides a global assessment of RNA quality. Although there may be large variability between the stability of different transcripts, most RNA transcripts should be appropriately represented when tissue is procured in an expedient manner [18].

Markers of expression can be reliably profiled in the clinical lab after appropriate measures have been taken to ensure high quality RNA for analysis. A panel of expression markers will best define the biology of a tumor and more accurately predict clinical behavior. Figure 4 compares the expression of *HER2* and *ER*

## 3A. Good RNA

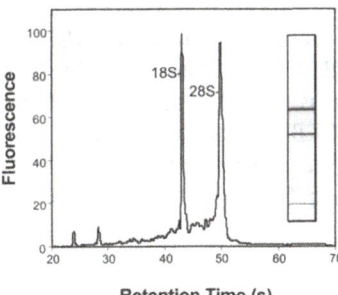

## 3B. Partially Degraded RNA

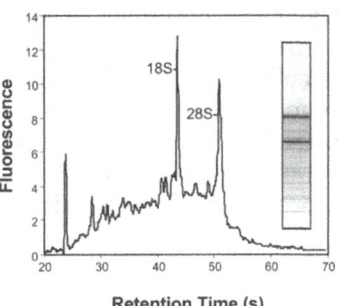

## 3C. Degraded RNA

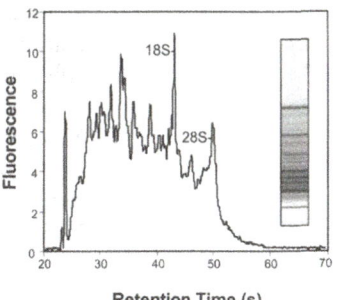

**Fig. 3.** Quality control for RNA. The intactness of RNA was evaluated using a capillary electrophoresis chip. The 18S and 28S band signals were compared to background signal using a chromatogram (fluorescence *versus* retention time) and gel-like representation (inset). The optimal ratio of 28S/18S should be 1.9–2.3. The marker dye (~25 sec) provides a scale for quantitative comparison. The first plot (**3A**) shows good RNA sample quality with very little degradation and distinct 18S and 28S ribosomal bands. Moderate degradation (**3B**) is reflected by overall lower fluorescence for the ribosomal peaks and higher background of degraded products. In a very poor quality sample (**3C**), the ribosomal bands are barely distinguished from the background of degraded RNA and may only be identified by the retention time

## 4A. *HER2* in MCF-7 and SKBR3

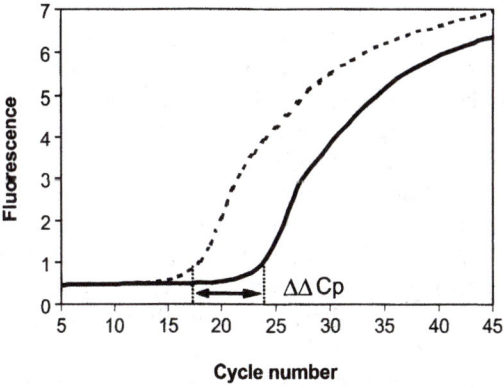

## 4B. *ER* in MCF-7 and SKBR3

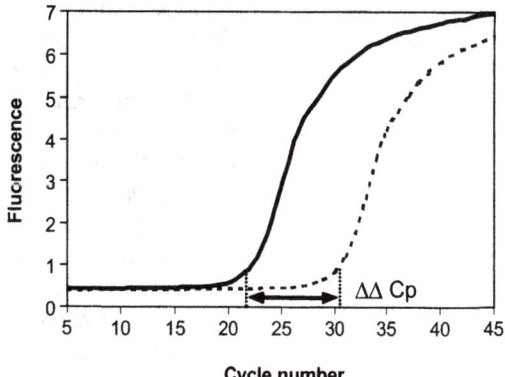

**Fig. 4.** Real-time quantitative PCR for gene expression analysis. Shown are PCR amplification curves for the predictive breast cancer markers *HER2* (**1A**) and *ER* (**1B**). The expression of each test gene was compared between the breast cell lines SKBR3 (– – – –) and MCF7 (——). The Cp (dotted line) of the test genes was determined using the 2nd derivative maximum method. In this example, two breast cell lines are compared rather than a normal breast sample. The control (*i.e.*, housekeeper) gene *PSMC4* was used as an indicator for the amount of cDNA added to each reaction

in two breast cancer cell lines. The SKBR3 cell line is strongly *HER2* positive compared to the MCF7 cell line (1A), whereas the converse is true for *ER* (1B). This is a rudimentary expression profile, however, it illustrates how select expression markers can make clear distinctions between tumor classifications. Using the efficiency of amplification for *HER2* (1.8) and *ER* (1.76) and the change in relative copy number for relative quantification, there is approximately 25-fold more *HER2* message and 90-fold less *ER* message for SKBR3 compared to MCF7. Using a ratio of the two test genes provides an even larger dynamic range for classification. For instance, the *HER2/ER* ratio is approximately 2000 for SKBR3 and 2 for MCF7.

## Comments

- Technology used today in cancer research, such as expression microarrays and CGH, allows genome-wide scanning and the discovery of altered genes involved in cancer. Microarrays have been used to identify coordinately expressed clusters of genes in breast cancer that predict clinical outcome. Statistical methods can be used to mine microarray data for finding control genes and smaller gene sets that are useful for molecular diagnostics. Real-time quantitative PCR is often used in research to validate microarray results and has potential as a clinical diagnostic technique.
- Real-time quantitative PCR using generic dsDNA dyes is a sensitive, accurate, and cost-effective method for determining RNA and DNA copy number. Profiling multiple genes within a sample improves diagnostic accuracy. Although sequence-specific probes offer internal controls due to multiplex capability, they require more optimization and become impractical when many targets are assayed at once.
- There are many software tools available for designing primers. Primers should be positioned properly within the gene for marker specificity (*e.g.*, RNA specific). Using primers with similar GC content and $T_m$ values allows many targets to be simultaneously assayed using the same PCR temperature parameters; thus allowing comprehensive profiling within a single run.
- The Cp value from the PCR amplification curve correlates to the starting amount of template in the reaction. The starting copy number for a gene can be estimated using an efficiency calibration curve. The relative copy number found between a test and a control gene assumes no differences in PCR efficiency or template input. The change in relative copy number allows differences in PCR efficiency and template input to be considered.
- Changes in DNA copy number are frequent events during the development and progression of breast cancer. These DNA alterations highly correlate with gene expression, raising the possibility that a DNA-based test could molecularly stratify breast cancers similar to the expression-based classifications. This has important implications for the development of clinical diagnostics because DNA-based classifications can be rapidly and reliably validated using archived tissue and readily implemented in the clinical laboratory.

# References

1. Chung CH, Bernard PS, Perou CM (2002) Molecular portraits and the family tree of cancer. Nat Genet 32 Suppl:533–540
2. Gancberg D, Lespagnard, L., Rouas, G., Paesmans, M., Piccart, M., DiLeo, A., Nogaret, J.M., Hertens, D., Verhest, A., Larsimont, D. (2000) Sensitivity of HER-2/neu antibodies in archival tissue samples of invasive breast carcinomas: Correlation with oncogene amplification in 160 cases. Am J Clin Pathol 113:675–682
3. Yarbro JW, Page DL, Fielding LP, Partridge EE, Murphy GP (1999) American Joint Committee on Cancer prognostic factors consensus conference. Cancer 86:2436–2446
4. Fisher B, Constantino, J.P., Wickerham, D.L., et al (1998) Tamoxifen for prevention of breast cancer: Report of the National Surgical Adjuvant Breast and Bowel Project P-1 Study. J Natl Cancer Inst 90:1371–1388
5. Slamon DJ, Leyland_Jones B, Shak S, Fuchs H, Paton V, Bajamonde A, Fleming T, Eiermann W, Wolter J, Pegram M, Baselga J, Norton L (2001) Use of chemotherapy plus a monoclonal antibody against HER2 for metastatic breast cancer that overexpresses HER2. N Engl J Med 344:783–792
6. Bilous M, Dowsett M, Hanna W, Isola J, Lebeau A, Moreno A, Penault-Llorca F, Ruschoff J, Tomasic G, Van De Vijver M (2003) Current perspectives on HER2 testing: a review of national testing guidelines. Mod Pathol 16:173–182
7. Tubbs RR, Pettay JD, Roche PC, Stoler MH, Jenkins RB, Grogan TM (2001) Discrepancies in clinical laboratory testing of eligibility for trastuzumab therapy: apparent immunohistochemical false-positives do not get the message. J Clin Oncol 19:2714–2721
8. Jacobs TW, Gown AM, Yaziji H, Barnes MJ, Schnitt SJ (1999) Comparison of fluorescence in situ hybridization and immunohistochemistry for the evaluation of HER-2/neu in breast cancer. J Clin Oncol 17:1974–1982
9. Pollack JR, Sorlie T, Perou CM, Rees CA, Jeffrey SS, Lonning PE, Tibshirani R, Botstein D, Borresen-Dale AL, Brown PO (2002) Microarray analysis reveals a major direct role of DNA copy number alteration in the transcriptional program of human breast tumors. Proc Natl Acad Sci U S A 99:12963–12968
10. De Preter K, Speleman F, Combaret V, Lunec J, Laureys G, Eussen BH, Francotte N, Board J, Pearson AD, De Paepe A, Van Roy N, Vandesompele J (2002) Quantification of MYCN, DDX1, and NAG gene copy number in neuroblastoma using a real-time quantitative PCR assay. Mod Pathol 15:159–166
11. Morrison TB, Weis JJ, Wittwer CT (1998) Quantification of low-copy transcripts by continuous SYBR Green I monitoring during amplification. Biotechniques 24:954–958, 960, 962
12. Rasmussen RP (2001) Quantification on the LightCycler. In: Wittwer CT, Meuer, S., Nakagawara, K. (ed) Rapid Cycle Real-Time PCR: Methods and Applications. Springer Verlag, Heidelberg, p 21–34
13. Pollack JR, Perou CM, Alizadeh AA, Eisen MB, Pergamenschikov A, Williams CF, Jeffrey SS, Botstein D, Brown PO (1999) Genome-wide analysis of DNA copy-number changes using cDNA microarrays. Nat Genet 23:41–46
14. Sorlie T, Tibshirani R, Parker J, Hastie T, Marron JS, Nobel A, Deng S, Johnsen H, Pesich R, Geisler S, Demeter J, Perou CM, Lonning PE, Brown PO, Borresen-Dale AL, Botstein D (2003) Repeated observation of breast tumor subtypes in independent gene expression data sets. Proc Natl Acad Sci U S A 100:8418–8423
15. Konigshoff M, Wilhelm J, Bohle RM, Pingoud A, Hahn M (2003) HER-2/neu gene copy number quantified by real-time PCR: comparison of gene amplification, heterozygosity, and immunohistochemical status in breast cancer tissue. Clin Chem 49:219–229
16. Perou CM, Sorlie T, Eisen MB, van de Rijn M, Jeffrey SS, Rees CA, Pollack JR, Ross DT, Johnsen H, Akslen LA, Fluge O, Pergamenschikov A, Williams C, Zhu SX, Lonning PE, Borresen-Dale AL, Brown PO, Botstein D (2000) Molecular portraits of human breast tumours. Nature 406:747–752

17. Kallioniemi OP, Kallioniemi A, Kurisu W, Thor A, Chen LC, Smith HS, Waldman FM, Pinkel D, Gray JW (1992) ERBB2 amplification in breast cancer analyzed by fluorescence in situ hybridization. Proc Natl Acad Sci U S A 89:5321–5325
18. Florell SR, Coffin CM, Holden JA, Zimmermann JW, Gerwels JW, Summers BK, Jones DA, Leachman SA (2001) Preservation of RNA for functional genomic studies: a multidisciplinary tumor bank protocol. Mod Pathol 14:116–128

# HER-2/neu Gene Copy Number Quantified by Real-Time PCR in Cell Lines and Breast Cancer Tissue

Melanie Königshoff, Jochen Wilhelm, Meinhard Hahn*

## Introduction

The *HER-2/neu* proto-oncogene, located at chromosomal region 17q21–22, encodes a 185 kDa transmembrane glycoprotein with tyrosine kinase activity, which plays an important role in signaling pathways of normal growth and development [1]. Oncogenic amplification has been observed in 20–30% of breast cancer cases [2], it predicts more frequent relapse and shorter survival times, and it affects response and resistance to therapy [3]. The promising treatment with humanized monoclonal antibody Herceptin against HER-2/neu receptor [4] calls for a reproducible and precise detection method for *HER-2/neu*.

Currently, the most common detection methods in routine clinical practice are immunohistochemical assays for *HER-2/neu* protein expression (HERCEPTest) [5], and fluorescence *in-situ* hydridization (FISH) tests for gene amplification [6]. However, PCR methods are likely to become more widely used because they are more sensitive, faster, and easier to perform [7]. Quantitative real-time PCR yields very precise results and does not need post-PCR analytical steps. In contrast to PCR methods for RNA [8,9], DNA-based assays must be able to discriminate very small concentration differences: the duplication of one allele represents a concentration difference of only 50%.

We developed a robust primer/probe system for rapid-cycle real-time PCR to precisely quantify *HER-2/neu* gene amplification. We also introduce for the first time *IGF-1* as a reference gene. This assay is well suited for clinical routine applications.

## Materials

LightCycler® instrument (Roche Diagnostics, Mannheim, Germany)    Equipment
LightCycler® capillaries (Roche Diagnostics, Mannheim, Germany)
LightCycler® centrifuge adaptors (Roche Diagnostics, Mannheim, Germany)
LightCycler® software, version 3.01 (Roche Diagnostics, Mannheim, Germany)

* Meinhard Hahn, Deutsches Krebsforschungszentrum, Abteilung Molekulare Genetik B060, Im Neuenheimer Feld 280, D-69120 Heidelberg, Germany, E-mail: m.hahn@dkfz.de

*So*FAR analysis software (own development) [10,11]

OLIGO Primer Analysis software, version 5.0 (National Biosciences, Plymouth, Minnesota USA)

U-3000 UV spectrophotometer (Hitachi/Colora, Lorch, Germany)

**Reagents**

QIAamp® DNA Blood reagent set (QIAGEN, Hilden, Germany)

10x PCR buffer, 100 mM Tris-HCl, 15 mM $MgCl_2$, 500 mM KCl, pH 8.3 at 20°C (Roche Diagnostics, Mannheim, Germany)

Bovine serum albumine, BSA; molecular biology grade (Roche Diagnostics, Mannheim, Germany)

*Taq* DNA polymerase, recombinant, 5 U/µl (Roche Diagnostics, Mannheim, Germany)

Deoxynucleotides, dATP, dCTP, dGTP, dTTP (Roche Diagnostics, Mannheim, Germany)

$MgCl_2$, 25 mM (Roche Diagnostics, Mannheim, Germany)

PCR primers, HPSF grade (MWG, Ebersberg, Germany)

Hybridization probes (TIB MOLBIOL, Berlin, Germany)

## Procedure

**Sample Preparation**

Excised breast cancer tissue samples from 51 women (age 41–86) with suspected breast cancer were analyzed. After routine embedding, the samples were snap-frozen in liquid nitrogen and stored at –80°C. Sections were stained with hematoxylin/eosin (H&E) to evaluate the tumor fraction. Samples containing more than 90% invasive tumor cells were used for further analysis. Fifteen 10 µm sections were used for genomic DNA preparation with the QIAamp DNA Blood reagent set.

Samples from the human breast cancer cell lines MDA-468, MCF-7, Cl.18, and SK-BR-3 were derived from ATCC stocks and DNA was extracted using the same reagent set as above.

After elution of the genomic DNA, the UV absorbance spectra of the solutions were recorded in the range of 220–320 nm. DNA concentrations were calculated using the double-strand DNA (dsDNA)-specific relation 1 $OD^{260\,nm}$ = 50 µg/ml. The DNA concentration of each sample was adjusted to 2 ng DNA/µl by diluting in water (3 ng were assumed to correspond to 1000 haploid genome equivalents).

**Oligonucleotides**

The sequences of the primers and probes used in this study are shown in Table 1. The length of the amplified sequence is 99 bp for *HER-/neu* and 90 bp for the reference gene *IGF-1*. All sequences were checked with the program Oligo 5.0 for absence of false priming sites, formation of primer dimers and primer/probe hybrids.

**Table 1.** Oligonucleotides

| *HER-2/neu* oncogene (GenBank Accession # M12036) | | | | | |
|---|---|---|---|---|---|
| | Position | Length | GC (%) | $T_m$ (°C) | Purity |
| **Primers** | | | | | |
| GAACTGGTGTATGCAGATTGC | 967–987 | 21 | 47.6 | 67.2 | 1.76 |
| AGCAAGAGTCCCCATCCTA | 1069–1051R | 19 | 52.6 | 67.6 | 1.82 |
| **Probes** | | | | | |
| GTATGCACCTGGGCTCTTTGC-AGGTCTCT-*F* | 992–1020 | 29 | 55.2 | 78.2 | 0.79 |
| *LCRed640*-CCGGAGCAAACCCC-TATGTCCACAAGG-*P* | 1021–1047 | 27 | 59.3 | 77.4 | 0.88 |
| *IGF-1* gene (GenBank Accession # M12659) | | | | | |
| **Primers** | | | | | |
| AGCTCGGCATAGTCTT | 1135–1150 | 16 | 50.0 | 66.1 | 1.97 |
| CCAAGTGAGGGGTGTGA | 1257R–1241 | 17 | 58.8 | 66.3 | 1.80 |
| **Probes** | | | | | |
| ATGAGACAGTGCCCTAAAGGGAC-CAATCCAATG-*F* | 1174–1206 | 33 | 48.5 | 75.7 | 0.85 |
| *LCRed640*-CTGCCTGCCCCTCCATA-GGTTCTAGGAAATGAG-*P* | 1207–1239 | 33 | 54.5 | 76.7 | 0.96 |

Before the PCR reaction mixtures were prepared, genomic DNA samples (2 ng/µl) were completely denatured by boiling in a water bath for 10 min [12] and a 2 µl-aliquot was transferred to the master mix.
The following master mix was used for amplification:

**LightCycler PCR**

**Table 2.** LightCyler PCR Master Mix Table

| Master Mix | Volume [µl] | [Final] |
|---|---|---|
| PCR buffer (10×) | 1 µl | 1× |
| BSA (5 mg/ml) | 1 µl | 0.5 g/L |
| Primers (5 µM each) | 1 µl | 0.5 µM each |
| dNTPs (2 mM each) | 1 µl | 0.2 mM each |
| MgCl₂ (25 mM) | 1.8 µl | 6 mM |
| Probes (2 µM each) | 1 µl | 0.2 µM each |
| *Taq* DNA polymerase (5 U/µl) | 0.1 µl | 0.5 U |
| H₂O | 1.1 µl | |
| Total master mix volume per reaction | 8 µl | |

Amplification protocol:

**Table 3.** Amplification Curve Analysis Parameter Table

| Parameter | Value | | |
|---|---|---|---|
| Cycles | 50 | | |
| Type | Amplification | | |
| | Segment 1 | Segment 2 | Segment 3 |
| Target temperature [°C] | 95 | 55 | 72 |
| Incubation time [s] | 1 | 10 | 5 |
| Temperature transition rate [°C/s] | 20 | 20 | 20 |
| Acquisition mode | None | Single | None |
| Gains | F1 = 5, F2 = 15 | | |

**Data Evaluation**

Raw data was analyzed with *SoFAR* analysis software [10,11].

The absolute target copy numbers were determined using a 1:2 dilution series of genomic DNA as an external standard. For the clinical samples, the gene copy numbers of *HER-2/neu* and *IGF-1* were determined in tumor and healthy control tissue. The relative copy number (Q) of *HER-2/neu* was calculated by:

$$Q = \frac{qT}{qN} = \frac{N^T_{HER\text{-}2/neu} \big/ N^T_{IGF\text{-}1}}{N^N_{HER\text{-}2/neu} \big/ N^N_{IGF\text{-}1}}$$

where qT is the ratio of *HER-2/neu versus IGF-1* copy numbers in tumor tissue; qN is the ratio of *HER-2/neu versus IGF-1* copy numbers in healthy control tissue; $N^T_{HER\text{-}2/neu}$ is the absolute copy number of *HER-2/neu* in tumor tissue; $N^N_{HER\text{-}2/neu}$ is the absolute copy number of *HER-2/neu* in healthy control tissue; $N^T_{IFG\text{-}1}$ is the absolute copy number of *IGF-1* in tumor tissue, and $N^N_{IFG\text{-}1}$ is the absolute copy number of *IGF-1* in healthy control tissue.

**Statistics**

The resolution of a factor of 2 was verified by a pairwise randomization test [13]. The probability p, calculated with this test, reveals the measured differences between the mean values of two data sets are either due to stochastic effects (p close to 1), or to real concentration differences of the analyzed targets (p close to 0).

To determine the linearity and reliability of the assay, simulation experiments of *HER-2/neu* gene amplification/deletion were performed using dilutions of blood lymphocyte DNA as *HER-2/neu* test samples. To simulate *HER-2/neu* gene amplification in the "tumor" sample, we used genomic DNA samples containing 5000, 2500, 500, and 250 copies of *HER-2/neu* (corresponding to 15, 7.5, 1.5, and 0.75 ng DNA). These samples were compared with a constant concentration of

*IGF-1* (always 500 copies per PCR). The ratios of *HER-2/neu* to *IGF-1* for "normal" samples were calculated from two independent reactions containing 500 copies per sample for *HER-2/neu* and *IGF-1*.

All samples were measured in triplicate. The mean $C_T$-values were used for the calculation of the Q values. Each Q value was determined for three different runs.

## Results

Based on the data bank entry (GenBank Accession No. M 12036), primers and probes were designed within the exon-2/intron-2 sequence of *HER-2/neu*. The standard PCR conditions for this system were optimized for a high signal-to-noise ratio by adjusting $Mg^{2+}$ and probe concentrations. Figure 1 shows amplification profiles and a calibration curve for *HER-2/neu*.

*Development of Primer and Probes Systems*

*IGF-1* proved to be a suitable reference gene to detect *HER-2/neu* amplification. Figure 2 shows amplification profiles and a calibration curve for *IGF-1*.

*Reproducibility.* The reproducibility of the assay was determined by statistical analysis of the quantification results for *HER-2/neu* and *IGF-1* copy numbers of a DNA sample (4 ng $\cong$ 1300 copies per reaction). Six replicates were analyzed in six repeated runs, on the same day for intra-run variation, and on three different days for inter-run variation.

*Statistical Analysis*

The intra-run variation for *HER-2/neu* varied between 7.6% and 17.5% (mean value: 11.8%; mean calculated copy number: 1357). The inter-run variations ranged between 10.9% and 23.4 % (mean value: 17.8%; mean calculated copy number: 1350). A pairwise randomization test, performed with nine aliquots of the same DNA sample (each measured four times), revealed a probability p of 0.998 (for both, *HER-2/neu* and *IGF-1*) that all samples have equal concentrations.

*Reliability.* The reliability was determined in simulation experiments for different amplification levels. The mean Q values were (mean ± SD): 9.8 ± 0.5 for 10-fold amplification, 4.9 ± 0.3 for five-fold amplification, 0.97 ± 0.06 for balanced sample, and 0.51 ± 0.06 for the deletion of one allele.

*Determination of cut-off values.* DNA samples of healthy volunteers contain the same copy numbers of test and reference genes (i.e., *HER-2/neu* and *IGF-1*). The Q value for two different DNA samples of the same individual is expected to be 1 (these two samples correspond to the DNA from tumor and healthy breast tissue of a patient in clinical studies). This was tested with blood lymphocyte DNA of healthy volunteers: The copy numbers of *HER-2/neu* and *IGF-1* were quantified in a 4 ng DNA sample. The mean Q value calculated from six experiments was 1.02 with a standard deviation of 0.06.

Based on these results, and for reasons of certainty, the Q cut-off value was set at 1.3 to be significant for amplifications and 0.7 for deletions.

The following cell lines were used: SK-BR-3 (high-level amplification of *HER-2/neu*), MDA-468 (low-level or no amplification), and MCF-7 Cl.18 (no amplifica-

*Analysis of Breast Cancer Cell Lines*

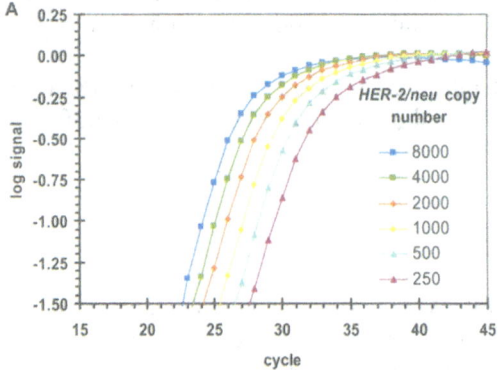

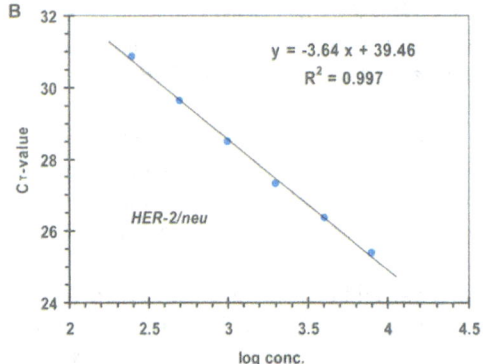

**Figure 1.** *HER-2/neu* amplification profiles and mean calibration curve. (**A**) *HER/2-neu* amplification profiles using hybridization probes for detection (Channel 1/ Channel 2). The DNA was diluted with water in steps of 1:2 from 8000 down to approximately 250 genomic equivalents (3 ng genomic DNA were assumed to be 1000 haploid genome equivalents). The negative controls without target DNA show no signal increase and are not visible in the diagrams. (**B**) *HER/2-neu* mean calibration curve (two-fold dilutions of human genomic DNA). The calibration curves, obtained by plotting log copy number *versus* $C_T$ and given in the form $y = a x + b$, were calculated using the fit-points method with arithmetic background correction

tion or allele deletion). The Q values, calculated using the mean $C_T$-values of the four-fold measured samples, were 9.0 for SK-BR-3, 0.8 for MDA-468, and 0.6 for MCF-7 Cl.18 (Figure 3) similar to prior data described in literature [14,15].

**Analysis of Clinical Samples**

Histologically, the tumor/tumor-like tissue samples from patients suspected to have breast cancer were classified as follows: 15 were non-neoplastic, 26 were invasive ductal carcinoma (IDC) of the breast, seven were invasive lobular breast cancer (ILC), along with one mucinous carcinoma, one tubular carcinoma, and one IDC-relapse.

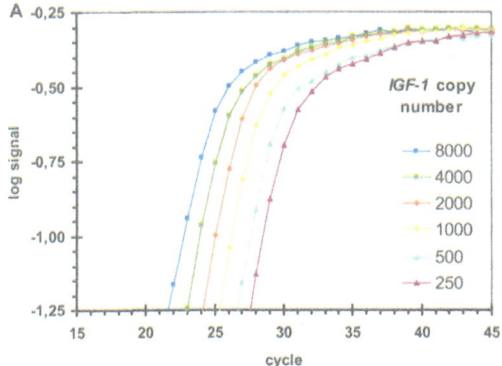

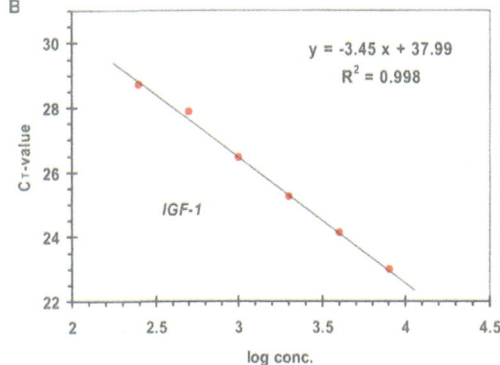

**Figure 2.** IFG-1 amplification profiles and mean calibration curve. (**A**) IGF-1 amplification profiles and (**B**) calibration curve using the same conditions shown in Figure 1

In 39 of 51 samples no gene amplification was detected (Q value: 1.02 ± 0.10), one sample showed a deletion (Q value: 0.51), and 11 samples showed a significant amplification with Q values varying between 1.41 and 11.24 (mean value: 2.87, median: 2.0). Figure 4 shows the amplification plots of a *HER-2/neu*-amplified breast cancer sample.

Amplifications and deletions of *HER-2/neu* were detected exclusively within the IDC samples.

Moreover, the IDC breast cancer samples were analyzed by immunohistochemistry (IHC) for protein overexpression. All immunohistochemical positive cases (n = 5) showed *HER-2/neu* gene amplification and were detected by rapid-cycle real-time PCR. In addition, six samples, not been identified by IHC, had significant levels of *HER-2/neu* amplification and were detected by quantitative real-time PCR.

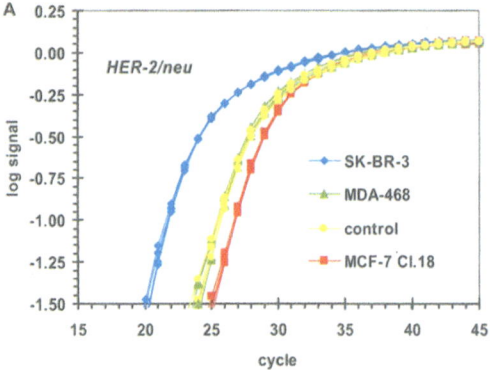

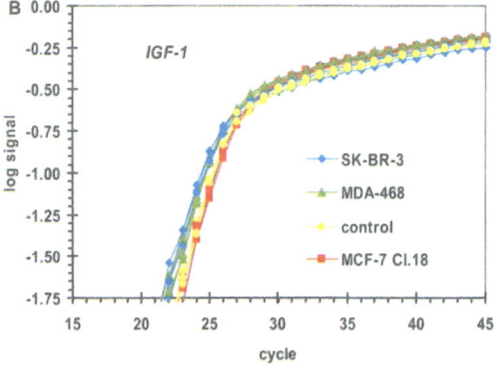

**Figure 3.** *HER-2/neu* and *IGF-1* quantification in three different human breast cancer cell lines. In each PCR experiment 4 ng genomic DNA of the specific cell line were analyzed in quadruplicate. In (**A**), *HER-2/neu*-specific amplification curves, in (**B**), *IGF-1*-specific curves are shown. Blood lymphocyte DNA of a healthy volunteer was applied for control purposes

## Comments

Good quality of the tumor tissue sample is crucial for exact quantification by real-time PCR using external standardization. Material was immediately shock-frozen in liquid nitrogen and stored at -80°C. Samples should be as homogeneous as possible and be checked by microscopy of thin sections. For real-time PCR, DNA amounts of 4 ng per 10 µl reaction are sufficient. This amount corresponds to ≈ 1330 copies of a single copy gene-like *HER-2/neu* or *IGF-1*. By using 4 ng DNA per PCR reaction, the scattering of results, due to limited template amounts, is negligible (< 3% for 1330 template copies *versus* > 30% for 100 template copies, data not shown). The analysis of larger amounts of DNA will not increase the accuracy significantly. At very large DNA concentrations due to complexation of free $Mg^{2+}$ amplification efficiency and quality of results can be reduced.

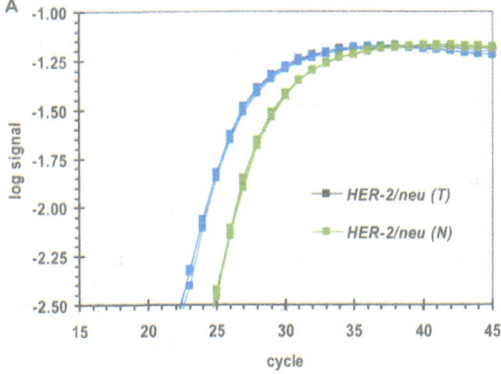

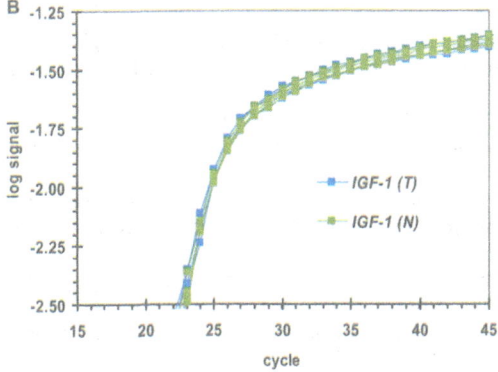

**Figure 4.** *HER-2/neu* and *IGF-1* quantification in tissue samples of a breast cancer patient. Homogenous tumor or healthy tissue areas were dissected under microscopic control from a series of consecutive 10 μm cuttings of the breast cancer sample. DNA was extracted from these pooled tissue dissections. In each PCR experiment, 4 ng genomic DNA of the specific sample (T = tumorous, N = normal breast tissue of the same patient) were analyzed in quadruplicate. In (**A**), the *HER-2/neu*-specific amplification curves are shown; in (**B**) the corresponding *IGF-1*-specific curves are shown. The calculated Q value for this sample, based on the quantification results, is 3.1, corresponding to a three-fold *HER-2/neu* gene amplification in the tumor tissue (*i.e.*, in mean six copies per cell)

IGF-1 was chosen as a reference gene [16], because the chromosomal region of this gene is known to be rarely involved in genomic alterations in breast cancer.

The *HER-2/neu* gene contains regions with high-melting domains, which can lead to unpredictable variations of $C_T$-values [12], and can make an assay unsuitable for precise and reliable quantitative analysis of gene copy numbers. The primer and probes used for this assay avoid high-melting domains of *HER-2/neu* and do not show position-dependent scattering of $C_T$-values. Similarly, no melting anomaly was observed in the case of the *IGF-1* gene.

# References

1. Yarden Y, Sliwkowski MX (2001) Untangling the ErbB signalling network. Nat Rev Mol Cell Biol 2:127–137
2. Slamon DJ, Clark GM, Wong SG, Levin WJ, Ullrich A, McGuire WL (1987) Human breast cancer: correlation of relapse and survival with amplification of the *HER-2/neu* oncogene. Science 235:177–182
3. Muss HB, Thor AD, Berry DA, Kute T, Liu ET, Koerner F, Cirrincione CT, Budman DR, Wood WC, Barcos M (1994) c-erbB-2 expression and response to adjuvant therapy in women with node-positive early breast cancer. N Engl J Med 330:1260–1266
4. Baselga J, Norton L, Albanell J, Kim YM, Mendelsohn J (1998) Recombinant humanized anti-HER2 antibody (Herceptin) enhances the antitumor activity of paclitaxel and doxorubicin against HER2/neu overexpressing human breast cancer xenografts. Cancer Res 58:2825–2831
5. Roche PC, Ingle JN (1999) Increased HER2 with U.S. Food and Drug Administration-approved antibody. J Clin Oncol 17:434–435
6. Pauletti G, Godolphin W, Press MF, Slamon DJ (1996) Detection and quantification of *HER-2/neu* gene amplification in human breast cancer archival material using fluorescence *in situ* hybridization. Oncogene 13:63–72
7. Wittwer CT, Ririe KM, Andrew RV, David DA, Gundry RA, Balis UJ (1997). The LightCycler: a microvolume multisample fluorimeter with rapid temperature control. BioTechniques 22:176–181
8. Specht K, Richter T, Müller U, Walch A, Werner M, Hofler H (2001) Quantitative gene expression analysis in microdissected archival formalin-fixed and paraffin-embedded tumor tissue. Am J Pathol 158:419–429
9. Bieche I, Onody P, Laurendeau I, Olivi M, Vidaud D, Lidereau R, Vidaud M (1999) Real-time reverse transcription-PCR assay for future management of ERBB2-based clinical applications. Clin Chem 45:1148–1156
10. Wilhelm J, Pingoud A, Hahn M (2003) *So*FAR: Software for fully automatic and highly accurate evaluation of real-time PCR data. BioTechniques 34:324–332
11. Wilhelm J, Pingoud A, Hahn M (2003) Validation of an algorithm for automatic quantification by real-time PCR. Anal Biochem 317:218–225
12. Wilhelm J, Hahn M, Pingoud A (2000) Influence of DNA target melting behavior on real-time PCR quantification. Clin Chem 46:1738–1743
13. Manly B (1997) Randomization bootstrap & Monte Carlo methods in biology. London: Chapman & Hall
14. Roetger A, Brandt B, Barnekow A (1997) Competitive-differential polymerase chain reaction for gene dosage estimation of erbB-1 (egfr), erbB-2, and erbB-3 oncogenes. DNA Cell Biol 16:443–448
15. Lyon E, Millson A, Lowery MC, Woods R, Wittwer CT (2001) Quantification of HER2/neu gene amplification by competitive PCR using fluorescent melting curve analysis. Clin Chem 47:844–851
16. Königshoff M, Wilhelm J, Bohle RM, Pingoud A, Hahn M (2003) *HER-2/neu* gene copy number quantified by real-time PCR: Comparison of gene amplification, heterozygosity, and immunohistochemical status in breast cancer tissue. Clin Chem 49:219–229

# Quantification of Several Components of Fibrinolytic and Matrix Metalloproteinase Systems in Primary Breast Cancer

Remedios Castelló, Francisco España, Justo Aznar, Amparo Estellés*

## Introduction

The fibrinolytic and matrix metalloproteinase (MMP) systems play a key role in the degradation of basement membrane and extracellular matrix (ECM), tissue remodeling, cancer cell invasion, and metastasis [1]. Urokinase plasminogen activator (uPA) catalyzes the conversion of plasminogen to plasmin, an enzyme that degrades a variety of ECM proteins, and activates MMPs and growth factors [2]. uPA activity is controlled by its principal inhibitor, the PA inhibitor type-1 (PAI-1) [3]. Tissue inhibitors of metalloproteinases (TIMPs), including TIMP-1 inhibit MMPs (collagenase, stromelysin and gelatinase) to regulate MMP activities [4].

In practice, breast tumor biopsies are usually small while conventional methods for RNA expression analysis, such as northern blotting, dot blot analysis, and RNase protection assay [5–7], require large amounts of RNA. Therefore, we used the LightCycler because of its high sensitivity and reproducibility [8] to develop assays to quantify PAI-1, uPA, and TIMP-1 mRNAs in breast cancer; β-actin was used to normalize the results.

Amplification of nonspecific PCR products, such as primer dimers, reduces sensitivity and can cause quantification PCR to fail. Hot start techniques [9] minimize the formation of undesired products. We used heat activation of the polymerase in the FastStart Reaction Mix SYBR Green I kit for a hot start to eliminate extra handling steps. To analyze the LightCycler data, we used the second derivative maximum method [10], which automatically calculates the crossing point cycle ($C_p$) values, reducing human error.

## Materials

Mastercyler® gradient (Eppendorff, Hamburg, Germany)   Equipment
LightCycler® instrument (Roche Diagnostics, Manheim, Germany)
LightCycler® capillary tubes (Roche Diagnostics, Manheim, Germany)
OLIGO Primer Analysis software, version 4.0

* Amparo Estellés, Hospital Universitario "La Fe", Centro de Investigación. Avda. Campanar 21, 46009 Valencia, Spain. E-mail: estelles_amp@gva.es

**Reagents**

Rneasy® Total RNA kit (Qiagen, Hilden, Germany)
SUPERSCRIPT™ RNase H⁻ Reverse Transcriptase (Gibco BRL, Life Technologies)
*High Pure* PCR™ Product Purification Kit (Roche Diagnostics, Manheim, Germany)
LightCycler – FastStart DNA Master SYBR Green I (Roche Diagnostics, Manheim, Germany)
Oligonucleotide primers (Invitrogen, Life Technologies).

## Procedure

**Sample Preparation**

Fifty-four patients (age 34–78 years, mean age 58 years) with primary, operable, unilateral breast cancer were studied. Tumor tissue was obtained during surgery and immediately snap-frozen in liquid nitrogen. Breast cancer tissue samples for our study were obtained in accordance with the Helsinki Declaration, and approved by the Ethical Committee of our Institution.

**RNA Extraction and cDNA Synthesis**

Total RNA and cDNA were obtained as previously described [11]. RNA from a frozen tumor was extracted with the RNeasy Total RNA kit, according to the manufacturer's instructions. One microgram of RNA was treated with DNase I (Invitrogen) and stored at –80°C. RNA concentration and purity were determined spectrophotometrically. One microgram of total RNA was reverse transcribed into first-strand cDNA using Superscript RNase H with oligo $(dT)_{15}$ priming (Promega). The cDNA was stored at –20°C.

**Primer Design**

Primers [11] were designed with the OLIGO Primer Analysis software and sequences were analyzed by FASTA in the EMBL database (http://www.embl-heidelberg.de) (Table 1).

**DNA External Standard Synthesis**

Quantification of uPA, PAI-1, and TIMP-1 transcripts were made with a DNA external standard curve generated by PCR. The target sequences were amplified using breast cancer tissue cDNA preparations and each primer pair. PCR amplifications were performed using the Mastercyler gradient. The PCR reaction mixture contained: 200 µM dNTPs, 2 mM $MgCl_2$, 0.5 µM each of either uPA, PAI-1, or TIMP-1 primers, or 0.25 µM of each β-actin primer, 1X PCR buffer (Roche), 1U Taq DNA Polymerase (Roche) and 5 µl of cDNA (1:10) in a final volume of 50 µl. The PCR conditions for uPA, PAI-1, and TIMP-1 (conditions for β-actin are given in parenthesis when different from the other targets) were: an initial 2-min denaturation step at 95 °C, followed by 35 cycles consisting of 45 s at 95°C (30 s at 95°C), 1 min at 60°C (30 s at 62°C), and 1 min at 72°C (45 s at 72°C), and ending with a 5-min elongation step at 72°C. Purification of the resulting cDNAs was done by column chromatography using the *High Pure* PCR Product Purification Kit and eluted with Tris-EDTA (pH=8.0) buffer. All samples had a unique band in agarose electrophoresis. The DNA amount was determined by PicoGreen fluorescence (Molecular Probes).

**Table 1.** Primers for the amplification of uPA, PAI-1, TIMP-1, and β-actin transcripts

| uPA (GenBank Accesion # BT007391) | | | | |
|---|---|---|---|---|
| | Position | Length | GC (%) | $T_m$ (°C) |
| CAC GCA AGG GGA GAT GAA | 741 | 18 | 55.6 | 61.8°C |
| ACA GCA TTT TGG TGG TGA CTT | 1081 R | 21 | 42.9 | 62.9°C |
| PCR product | | 341 | | |
| PAI-1 (GenBank Accesion # NM_000602) | | | | |
| TGC TGG TGA ATG CCC TCT ACT | 635 | 21 | 52.4 | 65.5°C |
| CGG TCA TTC CCA GGT TCT CTA | 1033 R | 21 | 52.4 | 63.2°C |
| PCR product | | 399 | | |
| TIMP-1 (GenBank Accesion # NM_003254) | | | | |
| CTG TTG TTG CTG TGG CTG ATA | 96 | 21 | 47.6 | 63.0°C |
| CCG TCC ACA AGC AAT GAG T | 576 R | 19 | 52.6 | 62.7°C |
| PCR product | | 481 | | |
| β-actin (GenBank Accesion # NM_001101) | | | | |
| CGT ACC ACT GGC ATC GTG AT | 512 | 20 | 55 | 63.4°C |
| GTG TTG GCG TAC AGG TCT TTG | 963 R | 21 | 52.4 | 62.7°C |
| PCR product | | 452 | | |

• The LightCycler analysis was performed in a reaction volume of 10 µl. A Master Mix of the following composition was used for uPA, PAI-1, and TIMP-1 (data for β-actin is given in parenthesis when different). **PCR with the LightCycler**

| | Volume [µl] | [Final] |
|---|---|---|
| LightCycler FastStart DNA Master SYBR Green I | 1 | 1 X |
| MgCl$_2$ (25mM) | 0.8 | 3 mM |
| Primers (5µM each) | 1 + 1 | 0.5 µM |
| | (0.6 + 0.6) | (0.3 µM) |
| H$_2$0 (PCR Grade) | 4.7 (5.5) | |

A total of 1.5 µl of cDNA sample (1:10) or standard was added to 8.5 µl of master mix in pre-cooled capillaries. Sealed capillaries were centrifuged (10 s at 2500 rpm) in a microcentriguge at 4°C and placed into the LightCycler rotor.

The following PCR protocols were used with SYBR Green I detection:
- Denaturation at 95°C for 10 min
- Amplification for uPA, PAI-1, and TIMP-1 (β-actin in parenthesis)

| Parameter | Value | | |
|---|---|---|---|
| Cycles | 50 | | |
| Type | Quantification | | |
| | Segment 1 | Segment 2 | Segment 3 |
| Target temperature [°C] | 95 | 60 (62) | 72 |
| Incubation time [s] | 15 | 5 | 18 (15) |
| Temperature transition rate [°C/s] | 20.0 | 20.0 | 20.0 |
| Acquisition mode | None | None | Single |

- Melting curve program

| Parameter | Value | | |
|---|---|---|---|
| Cycles | 1 | | |
| Type | Melting Curve | | |
| | Segment 1 | Segment 2 | Segment 3 |
| Target temperature [°C] | 95 | 65 | 95 |
| Incubation time [s] | 0 | 15 | 0 |
| Temperature transition rate [°C/s] | 20 | 20 | 0.1 |
| Acquisition mode | None | None | Single |

**Data Analysis**  The second derivative maximum method was used to determine the $C_p$ automatically for individual samples. The numbers of copies of unknown samples were calculated from the standard curve. Data were normalized using the ratio of the target cDNA concentration to that of β-actin. To verify the melting curve results, representative samples of the PCR products were assayed on 2% agarose gels. Two negative controls were included in each assay: one without template and one without reverse transcriptase.

## Results

The uPA, PAI-1, TIMP-1, and β-actin products produced melting curves with a sharp, single transition (Figure 1), indicating high purity and homogeneity of the PCR products. In a 2% agarose gel, only one band was observed in all samples. No band or signal was observed when either the template sample or the reverse transcription was omitted.

To evaluate the method's sensitivity, calibration curves were prepared for PAI-1, uPA, and TIMP-1 from known quantities of cDNA (10-fold dilutions from $10^6$ to 10 copies of cDNA per reaction). All calibration curves showed correlation

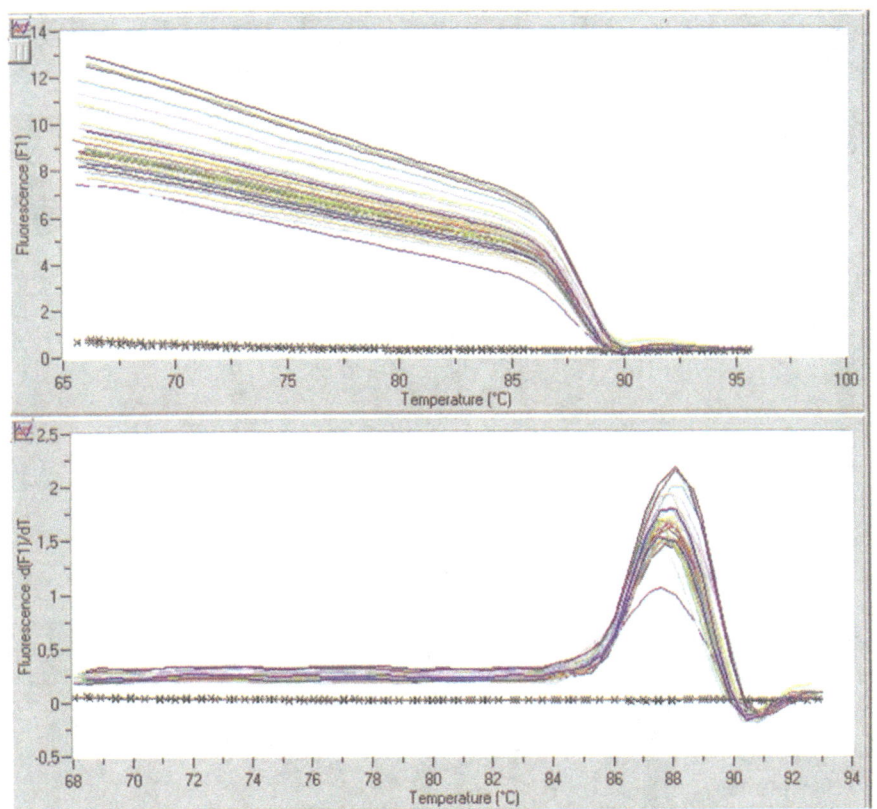

**Fig. 1.** Melting curves for all PCR products after uPA amplification. All controls and targets displayed one melting trasition. No product was observed when the template or the reverse transcriptase was omitted (-x-). PAI-1, TIMP-1, and β-actin showed similar reuslts

coefficients > 0.99, indicating a precise log-linear relationship. We observed the same detection limit (10 copies of cDNA) for the four target genes (Figure 2). The average slopes for the calibration curves were very similar: 3.68 for PAI-1, 3.58 for uPA, 3.66 for TIMP-1 and 3.65 for β-actin.

To test the reproducibility of PAI-1, uPA, TIMP-1 and β-actin mRNA quantification, intra- and inter-assay variation were determined from triplicate samples of the 11 targets and 5 controls. The difference in absolute $C_P$ values for each set of triplicates was < 0.57 cycles for intra-assay, and never >1.2 cycles for inter-assay.

These results indicate that quantitative real-time RT-PCR is a highly sensitive, reproducible, and fast method for measuring the gene expression of PAI-1, uPA, and TIMP-1 in breast cancer. When we applied this technique to breast cancer patients, we observed that the PAI-1 mRNA, uPA mRNA and TIMP-1 mRNA levels increased with TNM stage (Table 2). These components may be involved in breast cancer development and increased levels may be associated with a worse prognosis.

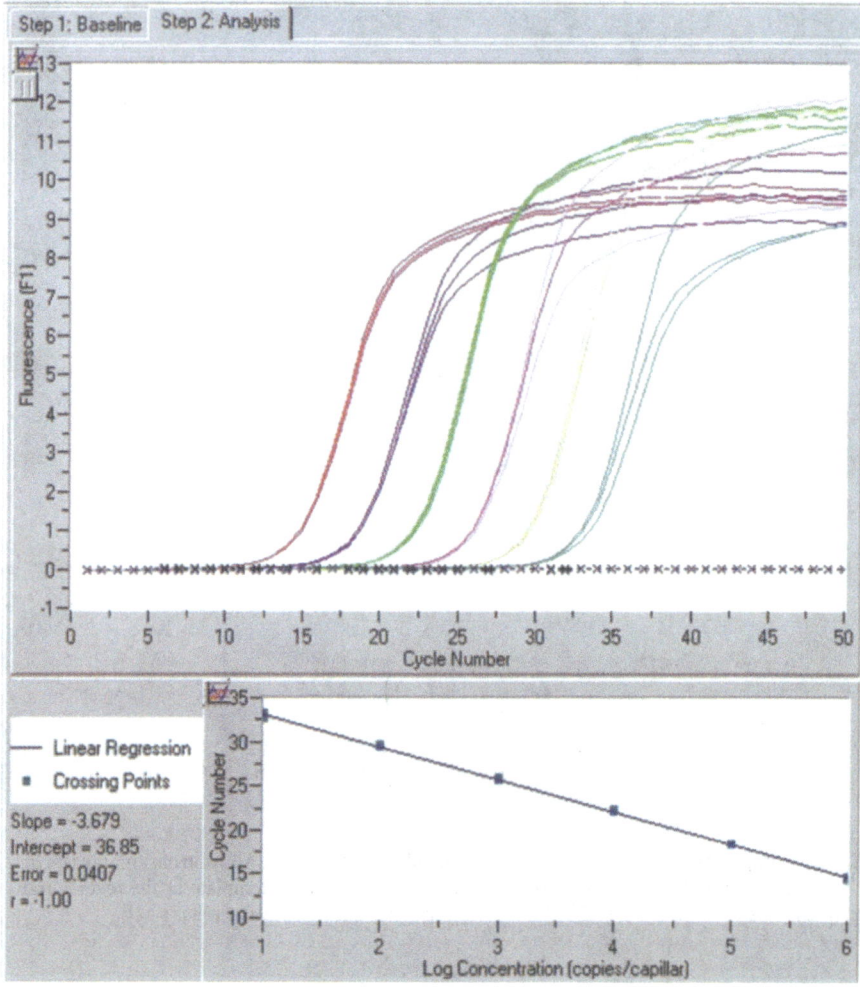

**Fig. 2.** Fluorescence data for PAI-1 standards together with the resulting calibration curve generated by the LightCycler software. PAI-1 diluted from $10^6$ to 10 copy numbers in triplicate. No product was amplified without template, or when reverse transcriptase was omitted (-x-). Similar results were obtained when PCR products of uPA and TIMP-1 were analyzed

## Comments

A critical parameter for specificity and efficiency of PCR is the formation of non-specific products before the first denaturation step. We have found that careful attention to primer design and the use of the LightCycler – FastStart DNA Master SYBR Green I Kit reduces the formation of undesired products.

**Table 2.** PAI-1, uPA and TIMP-1 antigen and mRNA levels and PAI-1 and uPA functional levels according to the grade of malignancy of breast cancer tissues

| | Grade I (n = 20) | Grade IIA (n = 21) | Grade IIB (n = 13) | Statistical Significance | | | |
|---|---|---|---|---|---|---|---|
| | | | | ANOVA | I-IIA | I-IIB | IIB-IIA |
| PAI-1 $_{mRNA}$ | $0.53 \pm 0.07$ | $0.81 \pm 0.15$ | $1.18 \pm 0.33$ | $p < 0.05$ | NS | $p < 0.05$ | NS |
| uPA$_{mRNA}$ | $1.92 \pm 0.16$ | $2.31 \pm 0.28$ | $3.30 \pm 0.69$ | $p < 0.05$ | NS | $p < 0.05$ | NS |
| TIMP-1 $_{mRNA}$ | $2.93 \pm 0.31$ | $3.37 \pm 0.44$ | $4.17 \pm 0.47$ | $p < 0.05$ | NS | $p < 0.05$ | NS |

Data are expressed as means $\pm$ SEM
mRNA values were normalized: mRNA = 100 x (target gene mRNA copies / ($\beta$-actin mRNA copies))
NS = not significant

In our study, a quantitative real-time RT-PCR method was developed to quantify PAI-1, uPA, and TIMP-1 mRNAs in breast cancer. This assay is a rapid, reproducible, and sensitive method to quantify gene expression. It requires smaller amounts of breast tissue than conventional methods. We believe that this quantitative real-time RT-PCR method will prove a valuable tool for further investigations on plasminogen activation and MMP systems in cancer and other diseases.

# References

1. Noël, A., Gilles, C., Bajou, K., Devy, L., Kebers, F., Lewalle, J. M., Maquoi, E., Munaut, C., Remacle, A., and Foidart, J. M. (1997) Emerging roles for proteinases in cancer. Invasion Metastasis 17: 221–239
2. Andreasen PA, Kjoller L, Christensen L, Duffy MJ. (1997) The urokinase-type plasminogen activator system in cancer metastasis. Int J Cancer 72: 1–22
3. Loskutoff DJ, van Mourik JA, Erickson LA, Lawrence D.(1983) Detection of an unusually stable fibrinolytic inhibitor produced by bovine endothelial cells. Proc Natl Acad Sci U S A 80: 2956–2960
4. Nagase H, Woessner JF Jr. (1999) Matrix Metalloproteinases. J Biol Chem 274: 21491–21494
5. Gabler C, Killian GJ, Einspanier R. (2001) Differential expression of extracellular matrix components in the bovine oviduct during the oestrous cycle. Reproduction 122:121–130
6. Estellés, A., Gilabert, J., Keeton, M., Eguchi, Y., Aznar, J., Grancha, S., España, F., Loskutoff, D. J., and Schleef, R. R. (1994) Altered expression of plasminogen activator inhibitor type 1 in placentas from pregnant women with preeclampsia and/or intrauterine fetal growth retardation. Blood 84: 143–150
7. Pacheco, M. M., Nishimoto, I. N., Mourao Neto, M., Mantovani, E. B., and Brentani, M. M. (2001) Prognostic significance of the combined expression of matrix metalloproteinase-9, urokinase type plasminogen activator and its receptor in breast cancer as measured by Northern blot analysis. Int. J. Biol. Markers16: 62–68
8. Morrison, T. B., Weis, J. J., and Wittwer, C. T. (1998) Quantification of low-copy transcripts by continuous SYBR Green I monitoring during amplification. Biotechniques 24: 954–958
9. Chou, Q., Russell, M., Birch, D. E., Raymond, J., and Bloch, W. (1992) Prevention of pre-PCR mis-priming and primer dimerization improves low-copy-number amplifications. Nucleic Acids Res 20: 1717–1723

10. Bernard PS, Wittwer CT. (2002) Real-time PCR technology for cancer diagnostics. Clin Chem 48: 1178–1185
11. Castelló, R., Estellés, A., Vazquez, C., Falco, C., España, F., Almenar, S. M., Fuster, C., and Aznar, J. (2002) Quantitative real-time reverse transcription-PCR assay for urokinase plasminogen activator, plasminogen activator inhibitor type 1, and tissue metalloproteinase inhibitor type 1 gene expressions in primary breast cancer. Clin Chem 48: 1288–1295

# A Rapid Cycle Real-Time Quantitative β-globin PCR for Quantification of Human DNA in Feces

CORNÉ H.W. KLAASSEN*, CLEMENS F.M. PRINSEN,
FREDERIK B.J.M. THUNNISSEN

## Introduction

At present, it is generally accepted that stool samples contain a certain amount of human DNA. This DNA originates from epithelial cells of the colonic mucosa and may contain information about the genetic status of the mucosal epithelium. The analysis of nuclear DNA extracted from stool specimens may have interesting clinically relevant applications, for instance in colorectal cancer screening procedures. This has already proven valuable in cancer diagnostics (1–3). Most studies published so far have focussed on the detection of sequence variations in tumor suppressor genes and oncogenes, and on correlation with clinical stage. However the ability to quantify the amount of human DNA in feces could also have clinically relevant applications. For instance, indirect evidence from the literature suggests that the amount of human DNA in feces from colorectal cancer patients may be elevated compared to healthy control individuals (4,5). Due to multiple technical barriers, direct quantification of human DNA from feces has gained little attention in the past. However, recent advances in real-time PCR procedures and DNA isolation protocols made it seem logical to use such an approach for quantitation of human DNA in stool samples as well. Therefore, we developed a rapid cycle real-time quantitative PCR assay targeting the human β-globin gene. This assay was used for direct and specific quantification of human DNA in stool samples.

## Materials

LightCycler (Roche Diagnostics, Almere, the Netherlands)    Equipment
LightCycler Capillaries (Roche Diagnostics)
LC Carousel Centrifuge (Roche Diagnostics)
DU 7500 spectrophotometer (Beckman)

* Corné H.W. Klaassen, Canisius Wilhelmina Hospital, Nijmegen, the Netherlands.
  Correspondence to Department of Pathology C66, Canisius Wilhelmina Hospital,
  Weg door Jonkerbos 100, 6532 SZ Nijmegen, the Netherlands.
  E-mail: c.klaassen@cwz.nl

| Software | LightCycler Software version 3 (Roche Diagnostics)<br>SRS6 (http://www.cmbi.kun.nl/srs6/)<br>BLAST (http://www.ncbi.nlm.nih.gov/BLAST/)<br>ClustalX (ftp://ftp-igbmc.u-strasbg.fr/pub/ClustalX/) |
|---|---|
| Kits | LightCycler FastStart DNA Hybridization Probes (Roche Diagnostics)<br>QIAamp DNA Stool Mini Kit (Qiagen, Hilden, Germany) |
| Reagents | PCR amplification primers (Eurogentec, Seraing, Belgium)<br>Hydrolysis probe for detection (Eurogentec)<br>Human placenta DNA (Sigma, St. Louis, USA) |

## Procedure

**Sample Preparation**

Fresh stool samples from healthy adult volunteers were processed within 48 hours after collection. Absolute care was taken to avoid hydration of the samples until further processing since this will result in rapid degradation of the DNA. DNA was extracted from 200 mg of stool using the QIAamp DNA Stool Mini Kit essentially according to the manufacturers guidelines for extraction of human DNA. Elution was in 200 µl. Total DNA yield and purity were determined by ultraviolet absorbance measurements, taking into account the conversion factor of 1 $OD_{260nm}$ ~ 50 µg/ml.

**Oligonucleotides**

Stool samples represent a complex mixture of microbial flora, food remains and human cells. Therefore, DNA extracted from stool will be a mixture of DNA's from various origins. A set of amplification primers and hydrolysis probe was designed based on DNA sequence alignments from β-globin sequences from edible species to ensure amplification and detection of human DNA only (figure 1). DNA sequences were retrieved from public databases using SRS6 software. Alignments were made using the ClustalX program (6). The specificity of primers and probe was verified by BLAST analysis (7). The sequences of primers and probe are given in table 1.

| Species | forward primer | Hydrolysis probe | reverse primer |
|---|---|---|---|
| Human | 5'-CTCCT**GGGCAACGTGCTGGTCTG**TGTGC**CTGGCCCATCACTTTGGCAAAGAA**TTCACCCC**ACCAGTGCAGGCTGCCT**ATCA-3' | | |
| Rabbit | ...................TAT.......T.T....T.................T.T.AG................ | | |
| Sheep | ...........T.....A..GGT.........T.GC..........G..........GGAGC.........AG.T... | | |
| Goat | ..................GGT.........T.GC...CA.....GT..........G.TGC.........A..T... | | |
| Pig | ...............A.A..GGT...T.....T.GC.G.C.....C.T..C.....A...GAAT...........T.T... | | |
| Cow | ...............A..GGT.....T.GCA.T.........G............GGTGC.........A..T... | | |
| Chicken | ........TG..A.C..CA..AT...C......GCC.....CA...G..C.....T..TGA.TGC.........GG.. | | |
| Duck | ........TG..A.C..CA..ATC..C......GCC.....CAC...G..T.....T..TGACTGC.....C....GG.. | | |

**Fig. 1.** Alignment of β-globin sequences from edible species and selected binding sites for amplification primers and detection probe. Dots represent residues identical to the human sequence. The following GenBank ID's were used: human, HSBGL3; Rabbit, OCBGLO; Sheep, OABCGLOB; Goat, CHHBBAA; Pig, SSBETAG; Cow, BTGL02; Chicken, GGGL02; Duck, CMBGA2B2

**Table 1.** Oligonucleotides

| Human β-globin gene (GenBank Accession # V00499) | | | | |
|---|---|---|---|---|
| | Position | Length | GC (%) | T$_m$ (°C) |
| Primers | | | | |
| GGGCAACGTGCTGGTCTG | 1454–1471 | 18 | 66.7 | 65.1 |
| AGGCAGCCTGCACTGGT | 1524R–1508 | 17 | 64.7 | 65.5 |
| PCR product | | 71 | | |
| Probe | | | | |
| FAM-CTGGCCCATCACTTTGGCAAAGAA-TAMRA | 1476–1499 | 24 | 50.0 | 67.4 |

## The following reaction mixes were prepared:

LightCycler PCR

- 6 µl of reaction mix and 14 µl of DNA (up to 500 ng) were added to a capillary. The capillaries were closed, placed into the LightCycler Carousel and spun once in the LC Carousel Centrifuge.

**Table 2.** Reaction mixes

| | Volume [µl] | [Final] |
|---|---|---|
| LC FastStart DNA Hybridization Probes (10x) | 2 | 1x |
| MgCl$_2$ (25 mM) | 1.6 | 3 mM |
| Amplification primers (10 µM each) | 0.5 + 0.5 | 250 nM |
| Detection probe (2 µM) | 1 | 100 nM |
| Water | 0.4 | – |
| Total volume | 6 | – |

## The following amplification protocol was applied:

- Denaturation at 95°C for 10 min
- Amplification

**Table 3.** Amplification parameters

| Parameter | Value | |
|---|---|---|
| Cycles | 45 | |
| Type | Quantification | |
| | Segment 1 | Segment 2 |
| Target temperature [°C] | 95 | 60 |
| Incubation time [s] | 0 | 15 |
| Temperature transition rate [°C/s] | 20 | 20 |
| Acquisition mode | None | Single |
| Gain settings | automatic | |

- Cooling at 40°C for 30 s

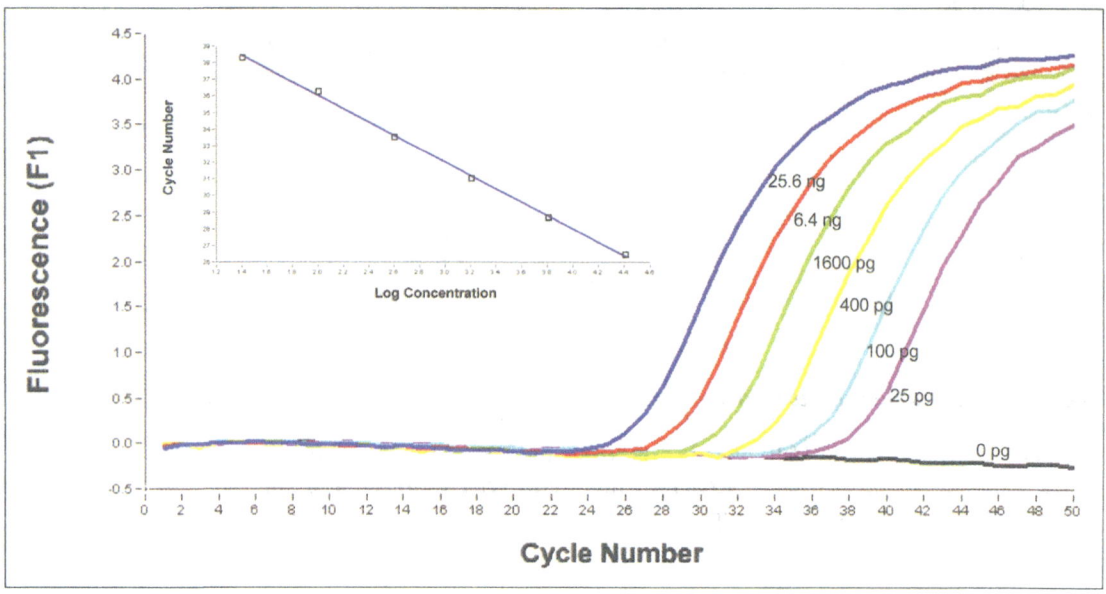

**Fig. 2.** Amplification plots and corresponding calibration curve (inset) obtained with human DNA standards

**Generation of Quantification Standards**

Quantification standards were created by making serial dilutions of human placenta DNA. The concentration of the stock solution was determined by ultraviolet absorbance measurements. Calibration curves were calculated from the obtained $C_T$ values in channel 1 using the second derivative maximum method.

## Results

**Quantification of Human DNA in Stool**

We report a real-time quantitative β-globin PCR. This assays was designed and optimized for specific quantification of human DNA in a complex background of DNA's from various origins. A detection limit of 30 pg of human DNA could be established which is equivalent to approximately only 5 human cells. DNA was extracted from stool samples from 15 healthy adult volunteers. The average amount of total DNA extracted from 200 mg of feces was ~20 µg. The $A_{260nm}/A_{280nm}$ was ~1.8 indicating an excellent purity of the DNA samples. Using the real-time quantitative PCR described here, the average amount of human DNA present in these samples was only 0.9 ng per µg of total DNA amounting to less than 0.1%.

**Absence of PCR Inhibitors**

Use of rapid cycle real-time PCR may result in undercalculation of the amount of human DNA present in the samples if they contain inhibitors of the real-time PCR process. The presence/absence of inhibitors was analyzed by spiking a total of 15 samples with a fixed amount of human DNA. When 500 ng of total DNA was analyzed, the average sample contained 0.45 ng of human DNA. When 500 ng of total DNA sample was spiked with 10 ng of human DNA, the average sample contained 10.3 ng of human DNA as determined by this real-time PCR assay (range: 7.8–16.6). So, from the 10 ng that was added, on average 9.85 ng (98.5 %) was recovered, demonstrating the virtual absence of PCR inhibitors and the accuracy of this quantification assay.

## Comments

Any human DNA extracted from feces is likely to have been subject to degradation. The chance of being able to PCR-amplify an amplicon of a certain size may be higher on degraded DNA if the amplicon is relatively small in size. The use of a hydrolysis probe enables the generation of smaller amplicons compared to the use of hybridization probes (standard chemistry hydrolysis probes are in general ~25 bp long versus a set of hybridization probes of ~50 bp). Furthermore, the use of a hydrolysis probe puts less stringent demands on the design of an assay compared to the use of a set of hybridization probes. For these reasons, we have chosen to design this quantitative assay based on a hydrolysis probe format. Clearly, the LightCycler system is able to deal with both assay formats although the use of a hydrolysis probe limits the possibilities of performing multiplexed assays.

*Hydrolysis Probe Versus Hybridization Probes*

According to our data, the average stool sample from healthy adult volunteers contains only 0.9 ng of human DNA per μg of total extracted DNA. This means that more than 99.9% of the extracted DNA is of other origin (most likely microbial). This may have important implications for studies on human DNA from stool samples. For instance, in colorectal cancer screening procedures where focus has been given to the detection of point mutations in human DNA from stool. If 1.0 μg of total fecal DNA is analyzed for the presence of mutations, extrapolation of our data indicates that this amounts on average to 0.9 ng of human DNA, which is the equivalent of only 150 human cells. In order to successfully detect mutations in this DNA, the ratio of tumor DNA relative to normal DNA should be more than 1:150. This calculation is similar to the results of Traverso and coworkers who demonstrated the ability to detect mutant APC genes from 14.1% down to 0.4% of the total pool of APC genes in DNA isolated from stool samples from colorectal cancer patients (8). In addition, recent evidence shows that the concentration of human DNA in stool samples from patients with distally located colorectal cancers may be elevated compared to controls (9). However, it has not yet been established that the surplus of human DNA that is present in these samples also originates from tumor cells. In all other colorectal cases, where the relative concentration of human DNA falls within the normal reference range, it would be beneficial to be able to modify the DNA isolation procedure in such a way that human DNA is preferentially purified over DNA from other origins. Some approaches to this effect have already been reported. For instance, Ahlquist et al. used gene-specific capture probes to enrich the DNA in specific sequences (5).

*Implications for Clinical Applications*

However, this approach also limits the number of human genes that can be studied. Therefore, other adaptations that enrich the final DNA sample in all human DNA sequences seem useful. The efficiency of these methods could very well be studied using the quantitative assay described here.

## References

1. Sidransky D, Tokino T, Hamilton SR, Kinzler KW, Levin B, Frost P and Vogelstein B. 1992 Identification of ras oncogene mutations in the stool of patients with curable colorectal tumors. Science 256:102–105
2. Caldas C, Hahn SA, Hruban RH, Redston MS, Yeo CJ and Kern SE. 1994 Detection of K-ras mutations in the stool of patients with pancreatic adenocarcinoma and pancreatic ductal hyperplasia. Cancer Res. 54:3568–3573
3. Tobi M, Luo FC and Ronai Z. 1994 Detection of K-ras mutation in colonic effluent samples from patients without evidence of colorectal carcinoma. J. Natl. Cancer Inst. 86:1007–1010
4. Villa E, Dugani A, Rebecchi AM, Vignoli A, Grottola A, Buttafoco P, Losi L, Perini M, Trande P, Merighi A, Lerose R and Manenti F. 1996 Identification of subjects at risk for colorectal carcinoma through a test based on K-ras determination in the stool. Gastroenterology 110:1346–1353
5. Ahlquist DA, Skoletsky JE, Boynton KA, Harrington JJ, Mahoney DW, Pierceall WE, Thibodeau SN and Shuber AP. 2000 Colorectal cancer screening by detection of altered human DNA in stool: feasibility of a multitarget assay panel. Gastroenterology 119:1219–1227
6. Thompson JD, Gibson TJ, Plewniak F, Jeanmougin F and Higgins DG. 1997 The CLUSTAL_X windows interface: flexible strategies for multiple sequence alignment aided by quality analysis tools. Nucleic Acids Res. 25:4876–4882
7. Altschul SF, Madden TL, Schäffer AA, Zhang J, Zhang Z, Miller W and Lipman DJ. 1997 Gapped BLAST and PSI-BLAST: a new generation of protein database search programs. Nucleic Acids Res. 25:3389–3402
8. Traverso G, Shuber A, Levin B, Johnson C, Olsson L, Schoetz Jr DJ, Hamilton SR, Boynton K, Kinzler KW and Vogelstein B. 2002 Detection of APC Mutations in Fecal DNA from Patients with Colorectal Tumors. N. Engl. J. Med. 346:311–320
9. Klaassen CHW, Jeunink MAF, Prinsen CFM, Ruers TJM, Tan ACITL, Strobbe LJA and Thunnissen FBJM. 2003. Quantitation of human DNA in feces as a diagnostic test for the presence of colorectal cancer. Clin. Chem. In the press

## Abbreviations

FAM is 6-carboxyfluorescein, TAMRA is 6-carboxytetramethylrhodamine.

# Quantification of Tumor Load for Follicular Lymphoma by Quantifying *BCL2/IGH* Using Real-Time Quantitative PCR by LightCycler™

CHUNG-CHE CHANG*, BERNARD C. SCHUR, B.S.

## Introduction

Here, we describe a newly developed real-time quantitative polymerase chain reaction (PCR) assay for detecting t(14;18)(q32;q21), the most frequent molecular lesion occurring in follicular lymphoma, using a rapid cycling format on a LightCycler™. Follicular lymphoma is one of the most frequent subtypes of malignant lymphoma in western countries and accounts for approximately 25% of all adult non-Hodgkin's lymphoma[1]. With current therapy the expected median survival is about 8–10 years [2]. More importantly, overall survival of FL patients has not been significantly improved in the last 2 decades despite different regimens using chemotherapy with/without radiation therapy [3–7].

The main application of this assay is to evaluate if the quantitation of tumor load, defined as the ratio of # of cells with t(14;18)/# of total cells, may predict which patients are at risk of clinical relapse/progression after treatment. The accurate assessment of this risk may improve the treatment efficacy and the outcomes of follicular lymphoma. One successful example is using RT-PCR to measure BCR-ABL fusion transcripts as a marker of tumor load in chronic myeloid leukemia (CML) [8–10]. For patients with CML, rising or persistently high levels of tumor load, BCR-ABL to ABL ratio of > 0.02% or > 100 BCR-ABL transcripts/μg RNA, in two sequential specimens more than 4 months following SCT are predictive of overt clinical relapse[8]. Furthermore, treating CML patients before there is overt clinical relapse of disease with donor lymphocytes infusion (DLI) has improved survival of CML patients after stem cell transplantation [9].

Our results indicate that the sensitivity, precision and linearity of this assay are favorably comparable to other recently reported real-time quantitative PCR assays for detecting t(14;18) [11–16]. These assays exploit the 5' exonuclease activity of Taq polymerase and measure PCR product accumulation as the reaction proceeds through dual-labeled fluorogenic probes. Furthermore, this assay uniquely provides an excellent turn-around time (40 minutes after the DNA extraction) that is much faster than other formats of real-time quantitative PCR assays currently in use. The rapid turn-around time can be particularly important

* Chung-Che Chang, M.D., Ph.D., Baylor College of Medicine, The Methodist Hospital,
  6565 Fannin, MS205, Houston, TX 77030
  E-mail: jeffchang@pol.net

in a clinical laboratory setting for prompt decision of patient management and efficient use of laboratory workforce.

## Materials

**Equipment**

Lambda 3A Spectrophotometer (Perkin-Elmer, Norwalk, CT, USA)
LightCycler™ (Roche Molecular Biochemical, Indianapolis, IN, USA)
Primer Designer, 4.0 (Sci-Ed Software, Durham, NC, USA)

**Reagents**

Puregene DNA Isolation Kit (Gentra Systems, Minneapolis, MN, USA)
Nucleotides (Roche Applied Science, Indianapolis, IN, USA)
Uracil-DNA Glycosylase, heat-labile (Roche Applied Science, Indianapolis, IN, USA)
LC Reaction Cuvettes (Roche Applied Science, Indianapolis, IN, USA)
AmpliTaq Gold (Applied Biosystems, Foster City, CA, USA)
Primers & Probes (IT Biochem, Salt Lake City, UT, USA)
Nuclease-Free Water (Ambion, Austin, TX, USA)
40mM & 50mM $Mg^{2+}$ +BSA (IT Biochem, Salt Lake City, UT, USA)

## Procedure

**Sample Preparation**

Total cellular DNA from the cell line SUDHL-6, containing t(14; 18)(q32;q21), normal human placental tissue, and patient samples was prepared using a Puregene DNA Isolation Kit. The amount of DNA extracted from SUDHL-6 cell line, placental tissue and patient sample cells was quantified spectrophotometrically. Extracted DNA purity was established by calculating the $A_{260}/A_{280}$ ratio. Quantification of SUDHL-6 DNA was further verified by limiting dilution PCR [17]. Patient DNA samples were stored at –20 °C while SUDHL-6 and placental DNA were processed as described below.

Tumor load is defined as the ratio of cells with t(14;18)(q32;q21) to the number of total cells (t(14;18)/total cells). Standardized samples for quantifying the tumor load were established by preparing five samples containing different ratios of SUDHL-6 DNA to total DNA. Four of these standardized samples contained tumor loads of 10% (1 tumor cell in 10 total cells), 1% (1 in 100), 0.1% (1 in 1,000), and 0.01% (1 in 10,000), respectively, and were used to establish a standard curve for quantitation. One additional standardized sample with a tumor load of 0.0025% (1 tumor cell in 40,000 total cells) was also prepared to test the sensitivity of the assay. All standardized samples were created by adding calculated amounts of placental DNA to calculated amounts of SUDHL-6 DNA to achieve the desired ratio while maintaining a total DNA concentration of 50 ng/ul. These samples were evaluated by real-time quantitative PCR as described below and once found acceptable were aliquotted to individual "Single Use Only" tubes and frozen at –86 °C.

Additionally, standardized samples for the reference gene, β-globin, were prepared. Total cellular DNA from normal human placental tissue was quantified spectrophotometrically. Extracted DNA purity was established by calculating the $A_{260}/A_{280}$ ratio was further verified by limiting dilution PCR [17]. A 50ng/ul aliquot of placental DNA was then serial diluted (1:10) to create the 4 standards (50ng/ul, 5ng/ul, 0.5ng/ul and 0.05ng/ul). These samples were evaluated and aliquotted as described above.

For quantitation of t(14; 18), the sequence of the 3' primer used was a consensus **Oligonucleotides** JH primer [18, 19]. Primer Designer 4.0 software was used to create the sequence of the 5' primer (Table 1). This primer was designed to allow the amplification of as small a DNA fragment as possible and still achieve optimal real-time PCR results when used with the consensus JH primer. The sequences of the probes (Table 1) were designed to be complimentary to the sense strand of the PCR products from bcl-2/JH translocation. The locations of the probes were designed to be 5' to all of the breakpoints most frequently observed in follicular lymphoma (Fig. 1). These sets of primer and probes are able to amplify and detect all bcl-2/JH rearrangements that occur at the major breakpoint region (MBR) as compared to the conventional PCR assay which was reported previously by us (data not shown) [20]. For quantitation of β-globin, primers and probes (Table 2) were designed using the Primer Designer software.

**Table 1.** Primers and probes used to quantify t(14;18)(q32;q21)

| GenBank accession number for bcl-2 is AY220759 and NG 001019 | | | | | |
|---|---|---|---|---|---|
| | Region | Length | GC (%) | $T_m$ (°C) | Purity |
| Primers (5' to 3') | | | | | |
| AGATGGCAAATGACCAGCAGA | 193432–52 | 21 | 47.6 | 67 | 1.981 |
| ACCTGAGGAGACGGTGACC | 960117–35 | 19 | 63.2 | 68 | 1.783 |
| Probes (5' to 3') | | | | | |
| LCRed640-CAGGCCACGTAAAGC-AACTC-P | 193482–501 | 20 | 55 | 66 | 1.102 |
| CTGGGTGGGTCTGTGTTGAA-F | 193503–522 | 20 | 55 | 66 | 1.068 |

**Table 2.** Primers and probes used to quantify β-globin

| GenBank accession number for β-globin is 455025 | | | | | |
|---|---|---|---|---|---|
| | Region | Length | GC (%) | $T_m$ (°C) | Purity |
| Primers (5' to 3') | | | | | |
| GAGACCATTGTGGCAGTGAT | 48517–536 | 20 | 50.0 | 65 | 1.772 |
| GGCTACGTTCATTGCCAGAT | 48660–679 | 20 | 50.0 | 68 | 1.680 |
| Probes (5' to 3') | | | | | |
| LCRed705-GCTGTCATCTCTGA-CTCTCTGTTTCACA-P | 48557–584 | 28 | 46.4 | 59 | 1.280 |
| TCCTTCTTTTCATTCACAGAC-AGGGATTTT-F | 48586–615 | 30 | 36.7 | 58 | 1.170 |

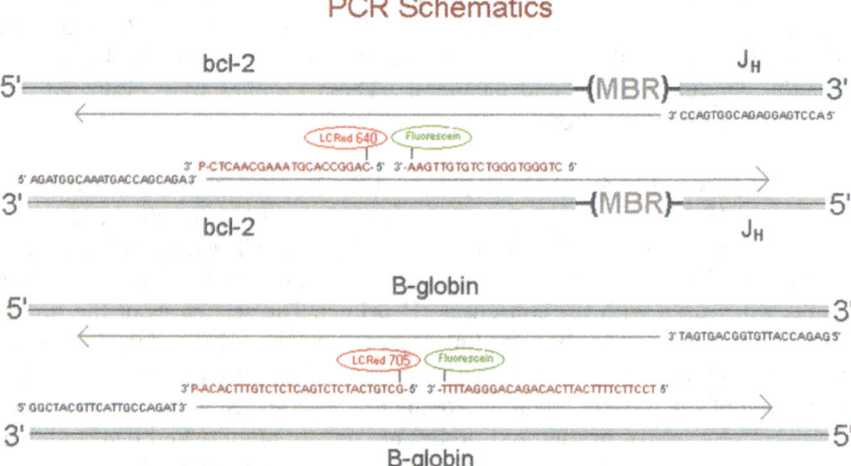

**Fig. 1.** Diagram of the locations of the primers and probes, relative to the major break point region (MBR) of t(14;18) and the β-globin gene

**Table 3.** Master mix used to quantify t(14;18)(q32;q21)

| Stock Solution | Volume [µl] | [Final] |
|---|---|---|
| Nuclease-Free Water | 1.8 | |
| MgCl$_2$ (45 mM) | 2.0 | 4.5 mM |
| BSA (2.5 mg/ml) – (included in MgCl$_2$) | – | 0.125 mg/ml |
| Nucleotides (1.25 mM) – (including dUTP in a 1:1:1:1:1 ratio) | 3.2 | 200 µM each |
| Primers (10 uM each) | 2.0 + 2.0 | 1.0 µM each |
| LCRed640 Probe (10 uM) | 0.4 | 0.2 µM |
| FITC Probe (10 uM) | 0.6 | 0.3 µM |
| Uracil-DNA Glycosylase (1 U/ul) | 1.0 | 1 U |
| AmpliTaq Gold (5 U/ul) | 2.0 | 10 U |
| Total master mix volume per reaction | 15.0 | |

The reaction mix detailed in Table 3 was aliquotted into LC Reaction Cuvettes. Next, 5 ul of each standardized sample with different tumor loads (1/10, 1/100, 1/1k, 1/10k, 1/40K) or patient sample containing 250 ng of total DNA was added for a total PCR reaction volume of 20 uL. The cuvettes were sealed, centrifuged in a micro-centrifuge and placed in the LightCycler™ rotor.

Using the following conditions, PCR amplification and quantification of PCR products were performed.

- Initial Denaturation and Activation of Amplitaq Gold Polymerase
Denaturation and activation of AmpliTaq Gold polymerase was accomplished by
a 10 min at 95 °C incubation.
- Amplification

**Table 4.** Amplification Parameters

| Parameter | Value | | | Value | | |
|---|---|---|---|---|---|---|
| Cycles | 5 | | | 55 | | |
| Type | Quantification | | | Quantification | | |
| Temp. Targets | Segment 1 | Segment 2 | Segment 3 | Segment 1 | Segment 2 | Segment 3 |
| Target Temperature [°C] | 95 | 67 | 74 | 95 | 63 | 74 |
| Incubation time [s] | 0 | 5 | 15 | 0 | 5 | 15 |
| Temp. Trans. Rate [°C/s] | 20 | 20 | 20 | 20 | 20 | 20 |
| Acquisition Mode | None | Single | None | None | Single | None |

- After PCR, the samples were cooled to 4 °C for 60 seconds.

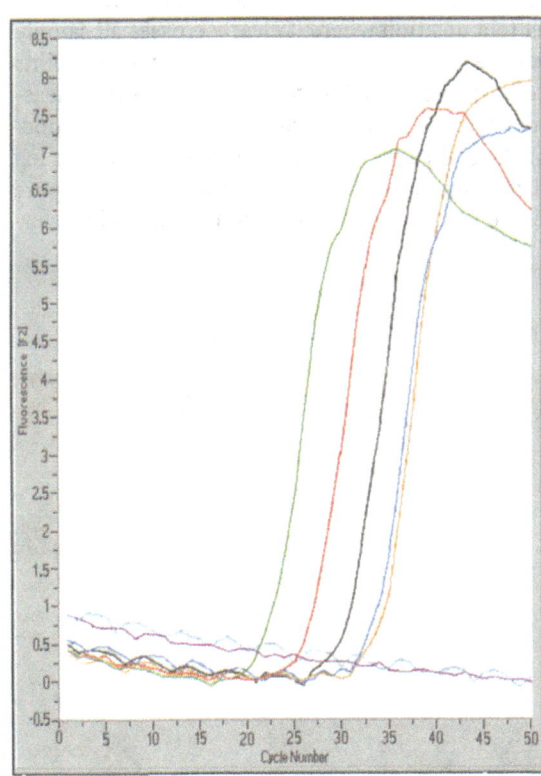

**Fig. 2.** Real-time monitoring of fluorescence intensity in each PCR cycle. The curves in green, red, black, and yellow represent standard samples with tumor load of 1/10, 1/100, 1/1K, and 1/10K respectively and the curves in pink, light blue and dark blue represents water (negative control), a patient without t(14;18) and a patient with t(14;18) respectively

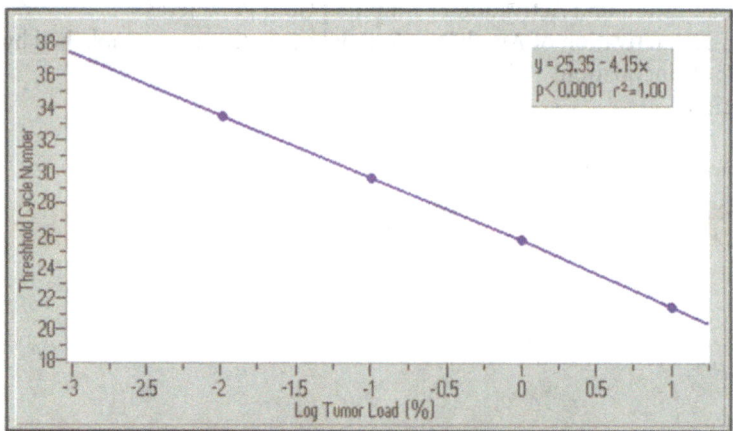

**Fig. 3.** The quantification standard curve was established by allowing the analysis software to calculate a best-fit line through the plots of the log of concentration (tumor load) in % (X-axis) versus the "threshold cycle number" (Y-axis)

Following PCR, a standard curve was established using the "Fit Points" analysis method in the LightCycler Data Analysis software using the standardized samples. Briefly, the noise band is set to a position where it clearly crosses all sample amplification curves in the lowest part of the log-linear phase. Three Fit Points along the linear phase of each reaction were used to get an accurate slope measurement. Crossing points were determined by using the "Minimize Error" function of the software. A best fit line through a plot of the crossing points versus log concentration for the 4 standardized samples is calculated by the software (Figure 2 and 3). The tumor loads of each patient sample were then calculated by referencing this standard curve.

**LightCycler β-Globin PCR**

**Table 5.** Master mix used to quantify β-globin

| Stock Solution | Volume [µl] | [Final] |
|---|---|---|
| Nuclease-Free Water | 3.0 | |
| MgCl₂ (45 mM) | 2.0 | 4.5 mM |
| BSA (2.5 mg/ml) – (included in MgCl₂) | – | 0.125 mg/ml |
| Nucleotides (1.25 mM) – (including dUTP in a 1:1:1:1:1 ratio) | 3.2 | 200 µM each |
| Primers (10 uM each) | 1.5 + 1.5 | 0.75 µM each |
| LCRed640 Probe (10 uM) | 0.8 | 0.4 µM |
| FITC Probe (10 uM) | 1.0 | 0.5 µM |
| Uracil-DNA Glycosylase (1 U/ul) | 1.0 | 1 U |
| AmpliTaq Gold (5 U/ul) | 1.0 | 5 U |
| Total master mix volume per reaction | 15.0 | |

The reaction mix detailed in Table 5 was aliquotted into LC Reaction Cuvettes. Next, 5 ul of each standardized sample with different amounts of DNA (250ng, 25ng, 2.5ng and 0.25ng) or patient sample containing 250 ng of total DNA was added for a total PCR reaction volume of 20 uL. The cuvettes were sealed, centrifuged in a micro-centrifuge and placed in the LightCycler™ rotor.

Using the following conditions, PCR amplification and quantification of PCR products were performed.

- Initial Denaturation and Activation of Amplitaq Gold Polymerase

Denaturation and activation of AmpliTaq Gold polymerase was accomplished by a 10 min, 95 °C incubation.

- Amplification

**Table 6.** Amplification Parameters

| Parameter | Value | | |
|---|---|---|---|
| Cycles | 40 | | |
| Type | Quantification | | |
| Temp. Targets | Segment 1 | Segment 2 | Segment 3 |
| Target temperature [°C] | 94 | 63 | 74 |
| Incubation time [s] | 0 | 5 | 15 |
| Temp. Trans. Rate [°C/s] | 20 | 20 | 20 |
| Acquisition Mode | None | Single | None |

- After PCR, the samples were cooled to 4 °C for 60 seconds.

Following PCR, a standard curve, using the data from the standardized samples, was established by the "Fit Point" analysis method in the LightCycler Data Analysis software as described in the LightCycler t(14;18) PCR section. The amount of DNA in each patient sample was then calculated by referencing this standard curve.

All the patients' samples were tested in duplicate initially for tumor load using the Bcl-2/IgH assay. If both results were negative, the sample was again tested in duplicate. Repeating a sample 4 times allowed us to reach a sensitivity of detection of one neoplastic cell in 160K normal cells. The tumor load was calculated as the average of the two or four runs.

For each patient sample, the amount of ß-globin gene was quantified separately using the RQ-PCR assay for ß-globin gene. This was to ensure that all samples contained similar amounts of DNA and were free of PCR inhibitors.

## Results

This RQ-PCR assay showed the sensitivity of detecting a single cell carrying t(14;18) in the background of 40,000 normal cells (i.e. a tumor load of 0.0025%). This is the highest sensitivity achievable with a total input of 250ng DNA (250 ng/6pg = 41700 cells, assuming 1 cell contains 6pg DNA). Further increasing the input of DNA to 500ng resulted in a decrease in PCR efficiency (data not shown). The linear measuring range of the standard curve of quantification of the assay was between tumor loads of 10% to 0.01% (Fig. 3). Additionally, reproducibility is demonstrated by an inter-assay (8 experiments over multiple days using the established standardized samples) coefficient of variation (CV) of $\leq$ 10.6% and an intra-assay (6 replicates of each established standardized sample within the same run) CV of $\leq$ 6.9%. The CV increased significantly when the tumor load was below 0.01% (data not shown).

Cryopreserved bone marrow and/or peripheral blood samples obtained at different time intervals after SCT from 11 patients (7 allogeneic, T-cell depleted/4 autologous) were tested for tumor load, as define by t(14;18) positive cells/total cells, using RQ-PCR. None of the 6 patients who remained in remission had samples with a tumor load > 0.01% after SCT, although fluctuating tumor loads of $\leq$ 0.01% were observed in three of these patients. In contrast, four of 5 patients (3 allogeneic/2 autologous) with relapsed/progressive disease had increasing tumor loads of > 0.01% after SCT (0/6 vs. 4/5, p < 0.02, Fisher Exact).

## Comments

It is important to verify the DNA quantity as measured by spectrophotometry with other methods, such as by limiting dilution assay. The results of spectrophotometric measurement were generally in agreement with the limiting dilution method for samples containing good quality DNA with a $A_{260}/A_{280}$ ratio greater than 1.7 (data not shown). Additionally, the presence or absence of PCR inhibitors must be monitored. The results of the β-globin assay indicated no PCR inhibitors were present in the tested samples (data not shown). These steps were essential for confirming the accuracy of standardized samples with different tumor loads.

This assay uses the standardized samples to closely mimic the clinical samples in which a variable number of tumor cells are present with normal cells. We report the tumor load as a ratio of neoplastic cells to total cells rather than absolute cell number to control for the minor differences in the amount of total input DNA. This approach together with verifying the DNA extraction efficiency by quantifying the ß-globin gene provides adequate control for variability in DNA isolation between patients' samples. This variability can be significant in clinical samples.

The standardized samples should be aliquotted and stored at a minimum of –20 °C. Degradation of DNA in samples was noted when they were stored at 4 °C or when they experienced repeated freezing and thawing processes, particularly for the samples with low tumor loads (data not shown).

In conclusion, our results indicate that this assay has excellent linearity, high sensitivity and good precision for the quantification of tumor load with a rapid turn around time of 30 to 40 minutes after DNA preparation. Our preliminary study suggested that a RQ-PCR measurable tumor load of >0.01% after stem cell transplantation for the treatment of follicular lymphoma may correlate with relapsed/progressive disease [23]. Prospective studies with greater numbers of cases are indicated to better determine the critical RQ-PCR tumor load that predicts poor outcome after SCT.

## References

1. Armitage JO and DD Weisenburger, (1998), New approach to classifying non-Hodgkin's lymphomas: clinical features of the major histologic subtypes. Non-Hodgkin's Lymphoma Classification Project. J Clin Oncol, 16:2780–95

2. Horning SJ, (1993), Natural history of and therapy for the indolent non-Hodgkin's lymphomas. Semin Oncol, 20:75–88

3. Rosenberg SA, (1985), Karnofsky memorial lecture. The low-grade non-Hodgkin's lymphomas: challenges and opportunities. J Clin Oncol, 3:299–310

4. Dana BW, S Dahlberg, BN Nathwani, E Chase, C Coltman, TP Miller, and RI Fisher, (1993), Long-term follow-up of patients with low-grade malignant lymphomas treated with doxorubicin-based chemotherapy or chemoimmunotherapy. J Clin Oncol, 11:644–51

5. Baldini L, A Guffanti, P Gobbi, M Colombi, M Federico, P Avanzini, L Cavanna, C Pieresca, V Silingardi, and AT Maiolo, (1997), A pilot study on the use of the ProMACE-CytaBOM regimen as a first-line treatment of advanced follicular non-Hodgkin's lymphoma. Gruppo Italiano per lo Studio dei Linfomi. Cancer, 79:1234–40

6. Hoppe RT, P Kushlan, HS Kaplan, SA Rosenberg, and BW Brown, (1981), The treatment of advanced stage favorable histology non-Hodgkin's lymphoma: a preliminary report of a randomized trial comparing single agent chemotherapy, combination chemotherapy, and whole body irradiation. Blood, 58:592–8

7. Federico M, U Vitolo, PL Zinzani, T Chisesi, V Clo, G Bellesi, M Magagnoli, M Liberati, C Boccomini, P Niscola, V Pavone, A Cuneo, G Santini, M Brugiatelli, L Baldini, L Rigacci, and L Resegotti, (2000), Prognosis of follicular lymphoma: a predictive model based on a retrospective analysis of 987 cases. Intergruppo Italiano Linfomi. Blood, 95:783–9

8. Lin F, F van Rhee, JM Goldman, and NC Cross, (1996), Kinetics of increasing BCR-ABL transcript numbers in chronic myeloid leukemia patients who relapse after bone marrow transplantation. Blood, 87:4473–8

9. Dazzi F, RM Szydlo, and JM Goldman, (1999), Donor lymphocyte infusions for relapse of chronic myeloid leukemia after allogeneic stem cell transplant: where we now stand. Exp Hematol, 27:1477–86

10. Bagg A, (2001), Commentary: minimal residual disease: how low do we go? Mol Diagn, 6:155–60

11. Mandigers CM, JP Meijerink, EJ Mensink, EL Tonnissen, KM Hebeda, MJ Bogman, and JM Raemaekers, (2001), Lack of correlation between numbers of circulating t(14;18)-positive cells and response to first-line treatment in follicular lymphoma. Blood, 98:940–4

12. Ladetto M, S Sametti, JW Donovan, D Ferrero, M Astolfi, M Mitterer, I Ricca, D Drandi, P Corradini, P Coser, A Pileri, JG Gribben, and C Tarella, (2001), A validated real-time quantitative PCR approach shows a correlation between tumor burden and successful ex vivo purging in follicular lymphoma patients. Exp Hematol, 29:183–93

13. Hirt C and G Dolken, (2000), Quantitative detection of t(14;18)-positive cells in patients with follicular lymphoma before and after autologous bone marrow transplantation. Bone Marrow Transplant, 25:419–26

14. Hosler GA, RO Bash, X Bai, V Jain, and RH Scheuermann, (1999), Development and validation of a quantitative polymerase chain reaction assay to evaluate minimal residual disease for T-cell acute lymphoblastic leukemia and follicular lymphoma. Am J Pathol, 154:1023–35

15. Dolken L, F Schuler, and G Dolken, (1998), Quantitative detection of t(14;18)-positive cells by real-time quantitative PCR using fluorogenic probes. Biotechniques, 25:1058–64

16. Luthra R, JA McBride, F Cabanillas, and A Sarris, (1998), Novel 5' exonuclease-based real-time PCR assay for the detection of t(14;18)(q32;q21) in patients with follicular lymphoma. Am J Pathol, 153:63–8

17. Brisco MJ, J Condon, E Hughes, SH Neoh, PJ Sykes, R Seshadri, I Toogood, K Waters, G Tauro, H Ekert, and et al., (1994), Outcome prediction in childhood acute lymphoblastic leukaemia by molecular quantification of residual disease at the end of induction. Lancet, 343:196–200

18. Slack DN, KP McCarthy, LM Wiedemann, and JP Sloane, (1993), Evaluation of sensitivity, specificity, and reproducibility of an optimized method for detecting clonal rearrangements of immunoglobulin and T-cell receptor genes in formalin-fixed, paraffin-embedded sections. Diagn Mol Pathol, 2:223–32

19. Bohling SD, TC King, CT Wittwer, and KS Elenitoba-Johnson, (1999), Rapid simultaneous amplification and detection of the MBR/JH chromosomal translocation by fluorescence melting curve analysis. Am J Pathol, 154:97–103

20. Juckett M, P Rowlings, M Hessner, C Keever-Taylor, W Burns, B Camitta, J Casper, WR Drobyski, G Hanson, M Horowitz, C Lawton, J Margolis, D Peitryga, and D Vesole, (1998), T cell-depleted allogeneic bone marrow transplantation for high-risk non-Hodgkin's lymphoma: clinical and molecular follow-up. Bone Marrow Transplant, 21:893–9

21. Wittwer CT, MG Herrmann, AA Moss, and RP Rasmussen, (1997), Continuous fluorescence monitoring of rapid cycle DNA amplification. Biotechniques, 22:130–1, 134–8

22. Wittwer CT, KM Ririe, RV Andrew, DA David, RA Gundry, and UJ Balis, (1997), The Light-Cycler: a microvolume multisample fluorimeter with rapid temperature control. Biotechniques, 22:176–81

23. Chang C, Bredeson C, Juckett M, Logan B, and Keever-Taylor CA, (2003), Tumor load in patients with follicular lymphoma post stem cell transplantation may correlate with clinical course. Bone Marrow Transplant, 32:287–291

# Real-Time Quantification of the *AML* Rearrangements, *AML1–ETO* and *TEL–AML1*, in Acute Leukemia

Eva Barragan, Pascual Bolufer*, Miguel Angel Sanz

## Introduction

Acute leukemia (AL) is a heterogeneous disease in its clinical presentation, its response to treatment, and especially in the genetic lesions associated with the development of the disease. The *AML1* gene, located on chromosome 21q22, encodes the $\alpha$ subunit of the core-binding factor (CBF) transcription factor [1]. *AML1* is one of the genes most frequently found in specific rearrangements characteristic of ALs. The *AML1* rearrangement is characterized by the t(8;21) (q21;q22) translocation that occurs in 7–10% of acute myeloid leukemias (AML) and 20–40% of AML-M2 [2]. The TEL-AML1 rearrangement is characterized by the t(12;21)(p13;q22) cryptic translocation, occurring in 20–30% of pediatric acute lymphoblastic leukemia (P-ALL) [3, 4], and 3% of adult ALL [5]. In all published studies, t(8;21) and t(12;21) are associated with relatively good prognoses compared with other ALs [2].

After initial treatment, most AL patients carrying *AML1* rearrangements achieve complete remission and the rearrangement becomes negative. However, a small fraction of these patients relapse [6]. Quantitative monitoring of specific chimeric transcripts in patients in complete remission may lead to the detection of minimal residual disease (MRD), which may predict the risk of hematological relapse before clinical manifestations become apparent.

The recent development of real-time PCR technology allows the levels of chimeric transcripts to be accurately quantified. Several methods for the quantification of *AML1–ETO* [7,8] and *TEL–AML1* [9,10] have been reported, but developed on relatively slow instruments using TaqMan probes. Fewer methods have been reported on the LightCycler system using hybridization-probe (HybProbe) chemistry [11,12].

One advantage of the LightCycler is high-speed thermal cycling that provides results in less than 45 min.

We report here two new methods for *AML1–ETO* and *TEL–AML1* quantification on the LightCycler using HybProbe chemistry. Results confirm that the proposed methods achieve the reliability demanded by clinical applications.

* Pascual Bolufer Gilabert, Laboratory of Molecular Biology, Department of Medical Biopathology, Escuela de Enfermería 7th floor, Hospital Universitario La Fe, Avda Capmanar 21, 46009 Valencia(Spain), E-mail: bolufer_pas@gva.es

## Materials

**Equipment**    LightCycler® instrument (Roche Diagnostics, Mannheim, Germany)
LightCycler® software, version 5.32 (Roche Diagnostics, Mannheim, Germany)

**Reagents**    TOPO TA Cloning® Kit (Invitrogen BV, Groningen, The Netherlands).
TaqMan® Reverse Transcription Kit (Applied Biosystems, Roche, Branchburg, New Jersey, USA).
LightCycler – FastStart DNA Master Hybridization Probes (Roche Molecular Biochemicals, Mannheim, Germany).
LightCycler® capillaries (Roche Molecular Biochemicals).
Heat-labile (1 U/µl) uracil–DNA–glycosylase (UDG; Roche Molecular Biochemicals).

## Procedure

**Patients**    *TEL–AML1* was quantitatively assessed in 27 samples (21 bone marrow [BM] and 6 peripheral-blood [PB] samples) from 10 P-ALL positive samples for this rearrangement. The samples were collected at the time of active disease and during follow-up.

*AML1–ETO* was quantified in a total of 23 BM samples from eight *AML-M2* positive patients, collected at diagnosis, and during follow-up.

**Sample Preparation**    Mononuclear cells were isolated from EDTA-anticoagulated PB by Lymphoprep (Nycomed, Pharma AS) density-gradient centrifugation. The cells collected
**RNA Isolation**    (mean number $5 \times 10^6$) were resuspended in guanidinium thiocyanate solution (4 M guanidinium thiocyanate, 25 mM sodium citrate, [pH 7], containing 5 g/l sarcosyl and 0.1 M 2-mercaptoethanol) and stored at –80°C until RNA extraction. RNA was extracted using the guanidinium thiocyanate–phenol–chloroform procedure of Chomczynski and Sacchi [13].

**cDNA Synthesis**    RNA (0.5 µg) was transcribed to cDNA in a 25-µl reaction volume using the TaqMan Reverse Transcription Kit with random hexamer primers following the manufacturer's instructions. The reverse transcription program consisted of incubations for 10 min at 25°C, 30 min at 48°C, and 5 min at 95°C.

**Standard Curves**    Standard curves were prepared using a pCR II-TOPO vector into which the PCR products *TEL–AML1* (pCR II-TOPO[TEL-AML1]), *AML1–ETO* (pCR II-TOPO[AML1-ETO]), or *AML1* (pCR II-TOPO[AML1]) were cloned according to the manufacturer's instructions (TOPO Cloning Kit). PCR products were amplified from cDNA samples of positive patients using the primers and conditions described by Satake *et al.* [14] for *TEL–AML1* or by Kozu *et al.* [15] for *AML1–ETO* and *AML1*.

The standard curves were prepared using 10-fold serial dilutions of the plasmids pCR II-TOPO[TEL-AML1], pCR II-TOPO[AML1-ETO], or pCR II-TOPO[AML1] in heterologous salmon sperm DNA (40 µg DNA/mL). The *TEL–AML1* standard curve consisted of seven points, ranging from $4.5\times10^6$ to 4.5 copies. *AML1–ETO* and *AML1* standard curves consisted of six points, ranging from $22\times10^6$ to 22 copies, and from $10\times10^6$ to 10 copies, respectively.

Although two transcripts for *TEL–AML1* have been reported, the long form (L form) and the short form (S form), we only analyzed the L form, because it is the predominant form in more than 90% of P-ALL. In this form, the *TEL* gene fuses at nucleotide (nt) 1033 (exon 5) to nt 542 of *AML* (exon 3) [14,16,17].

Oligonucleotides

TEL–AML1
Oligonuceoltides

To amplify *TEL–AML1* transcripts, we used the forward primer of Satake *et al.* [14], which binds the *TEL* gene, together with a reverse primer that localizes to *AML* exon 2 (Table 1 and Figure 1). This set of primers is specific to the long transcript (L form), because the PCR product requires the presence of exon 2 [18,19]. The 5' and 3' fluorescently-labeled HybProbes (TIB MOLBIOL, Berlin, Germany) hybridize to the amplified *TEL* gene.

To detect *AML1–ETO* transcripts, we used the primers of Satake *et al.* [20]. The 5' and 3' fluorescently-labeled HybProbes (TIB MOLBIOL) hybridize to regions in the amplified *ETO* fragment.

AML1–ETO
Oligonucleotides

*AML1* was amplified as the control gene using forward and reverse primers described by Kozu *et al.* [15] with 5' and 3' fluorescently-labeled HybProbes (TIB MOLBIOL) (Table 1). Both primers and probes hybridize in *AML1* exon 5 [12].

AML1
Oligonucleotides

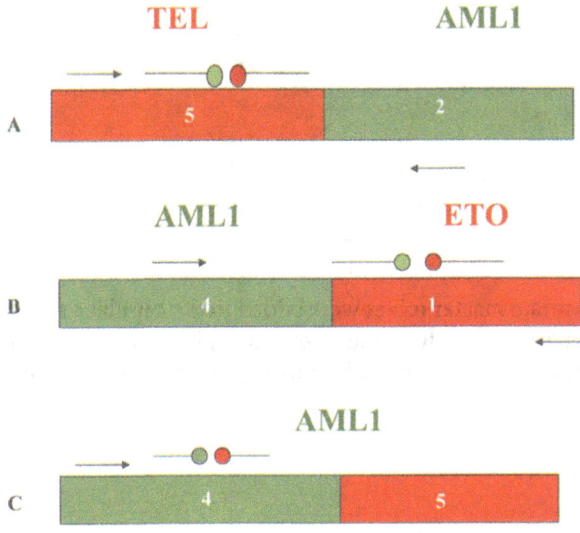

**Fig. 1. (A)** Schematic diagram of the *TEL–AML1* transcript with *TEL* exon 5 fused to *AML1* exon 2. **(B)** Schematic diagram of the *AML1–ETO* transcript with *AML1* exon 4 fused to *ETO* exon 1. **(C)** Schematic representation of the *AML1* control gene. The positions of the forward and reverse primers (arrows) and hybridization probes (colored filled circles followed by tails) are indicated

**Table 1.** Oligonucleotides

| TEL (GenBank Accession # U11732)-AML1 (GenBank Accession #D43969) | | | | | |
|---|---|---|---|---|---|
| | | Position | Length | GC(%) | $T_m$ (°C) |
| **Primers** | | | | | |
| TEL2 (+) | AACCTCTCTCCATCGGGAAGA | 937–956 | 21 | 52.4 | 67.5 |
| X2AML(-) | GGACGTCTCTAGAAGGATTCAT | 538–517 | 22 | 45.5 | 63.9 |
| **Probes** | | | | | |
| TEL 3FL | TGGTCTCTGTCTCCCCGCCTG-AAGA F | 580–604 | 25 | 60.0 | 75.2 |
| TEL 5LC | LC RED640-ACGCCATGCCCAT-TGGGAGAATAGC-P | 607–630 | 25 | 56.0 | 74.0 |
| **AML1 (GenBank Accession #D43969)-ETO (GenBank Accession #D14289)** | | | | | |
| **Primers** | | | | | |
| AML1F (+) | CACTGTCTTCACAAACCCAC | 975–994 | 20 | 50 | 61.2 |
| MTJ3 (-) | ATGAACTGGTTCTTGGAGCT-CCTT | 374–397 | 24 | 45.8 | 65.8 |
| **Probes** | | | | | |
| MTG 3FL | AGACGCAATCTAGGCTGACT-CCTCCA-F | 323–348 | 26 | 53.8 | 69.7 |
| MTG 5LC | LC RED640-CAATGCCACCTC-CCCCAACTACTC-P | 350–373 | 24 | 58.3 | 68.1 |
| **AML1 (GenBank Accession #D43969)** | | | | | |
| **Primers** | | | | | |
| AML1C (+) | GAGGGAAAAGCTTCACTCTG | 950–969 | 20 | 50.0 | 63.4 |
| AML1E (-) | GCCGCAGCTGCTCCAGTTCA | 1141–1122 | 20 | 65.0 | 73.2 |
| **Probes** | | | | | |
| AML1 3FL | AACCCACCGCAAGTCGCCA-CCT- F | 988–1109 | 22 | 63.6 | 76.1 |
| AML1 5LC | LC RED640-CCACAGAGCCA-TCAAAATCACAGTGGATGG-P | 1011–1040 | 30 | 50.0 | 73.6 |

**LightCycler PCR**  Aliquots of 8 µl of the appropriate master mixes were added to the capillary tubes followed by 2 µl of cDNA corresponding to the unknown samples or to points on the standard curves. The capillaries were capped, centrifuged, and placed in the LightCycler rotor.

**Table 2.** Master Mix

| TEL-AML1 Master Mix | Volume [µl] | [Final] |
|---|---|---|
| FastStart LightCycler – DNA Master Hybridization Probes | 1.0 | 1x |
| $MgCl_2$ (25 mM) | 1.6 | 5 mM |
| Mix of primers M1 or M2 (5 µM both) | 0.8 | 0.4 µM both |
| Hybridization Probes (4 µM each) | 0.5+0.5 | 0.2 µM each |
| Heat-Labile Uracil DNA Glycosilase (1 U /ml) | 0.5 | 0.5 U |
| $H_2O$ (PCR grade) | 3.1 | |
| cDNA samples or standards | 2.0 | |
| **Total Volume** | **10.0** | |
| **AML1-ETO & AML1 Master Mix** | **Volume [µl]** | **[Final]** |
| FastStart LightCycler™-DNA Master Hybridization Probes | 1.0 | 1x |
| $MgCl_2$ (25 mM) | 0.8 | 3 mM |
| Primers (5 µM each) | 1+1 | 0.5 µM |
| Hybridization Probes (4 µM each) | 0.5+0.5 | 0.2 µM |
| Heat-Labile Uracil DNA Glycsolilase (1 U /µl) | 0.5 | 0.5 U |
| $H_2O$ (PCR grade) | 2.7 | |
| cDNA samples or standard points | 2.0 | |
| **Total Volume** | **10.0** | |

The following protocols were used for amplifications:                                      **PCR protocols**

**Table 3.** PCR protocols

| TEL-AML1 Parameters | Program 1 | Program 2 | Program 3 | | |
|---|---|---|---|---|---|
| Cycles | 1 | 1 | 45 | | |
| Type | UDG | Denaturation | Amplification | | |
| | | | Seg 1 | Seg 2 | Seg 3 |
| Target temperature [°C] | 32 | 95 | 96 | 61 | 72 |
| Incubation time [s] | 300 | 600 | 2 | 10 | 10 |
| Temperature transition rate [°C/s] | 20 | 20 | 20 | 20 | 2 |
| Acquisition mode | None | None | None | Single | None |
| Channels | | | | F2/F1 | |
| **AML1-ETO & AML1 Parameters** | **Program 1** | **Program 2** | **Program 3** | | |
| Cycles | 1 | 1 | 45 | | |
| Type | UDG | Denaturation | Amplification | | |
| | | | Seg 1 | Seg 2 | Seg 3 |
| Target temperature [°C] | 32 | 95 | 94 | 60 | 72 |
| Incubation time [s] | 300 | 600 | 0 | 12 | 12 |
| Temperature transition rate [°C/s] | 20 | 20 | 20 | 20 | 4 |
| Acquisition mode | None | None | None | Single | None |
| Channels | | | | F2/F1 | |

**Quantification**

Quantification was performed using LightCycler software, version 5.32, with the "second derivative maximum" option. Unknown concentrations were calculated from the regression line of the standard curve (crossing point [$C_p$] vs. the logarithm of the concentration) Quantified levels of *TEL–AML1* and *AML1–ETO* were normalized relative to the expression of the control gene, *AML1*. The final results were expressed as the ratio of the number of copies of the fusion transcripts to the number of copies of the *AML1* transcript.

## Results

**Study of the Method**

**Quantification Standard Curves**

The regression coefficients of five consecutive *TEL–AML1* standard curves were all –1.0. The mean ± standard deviation (SD) of the slope was –3.26 ± 0.15, and 39.16 ± 1.09 for the intercept. The mean $C_p$ on five consecutive TEL–AML1 standard curves was almost constant with an SD < 0.7 cycles for all the points, except for the lowest point (five plasmid copies), which had an SD of 1.28 cycles.

The regression coefficients for five *AML1–ETO* standard curves were all greater than 0.99. The mean ± SD of the slope was of –3.75 ± 0.21, and the intercept 38.74 ± 0.78.

The five *AML* standard curves, all with regression coefficients > 0.99, had a slope of –4.26 ± 0.26, and an intercept of 37.42 ± 1.06.

**PCR Efficiency**

The mean PCR efficiency ± SD, calculated from the slope of the standard curves ($10^{-1/slope}$) was 2.00 ± 0.08 for *TEL–AML1*, 1.83 ± 0.06 for *AML1–ETO*, and 1.692 ± 0.05 for *AML1*.

**Quality of Amplified Products**

The PCR products amplified for *TEL–AML1* were removed from the capillaries and analyzed by electrophoresis on 2% agarose minigel, verifying the absence of artifacts and the expected size of the amplified products (134 bp).

We also observed an absence of artifacts in the amplification of *AML1–ETO* and *AML1*, and a good correspondence between the sizes of the PCR products and their expected sizes (213 bp for *AML1–ETO* and 192 bp for *AML1*).

**Sensitivity Studies**

The PCR sensitivity of *TEL–AML1* and amplification was assessed by analyzing a series of 10-fold aqueous dilutions (range: $10^{-2}$ to $10^{-5}$) of a *TEL–AML1*-positive cDNA sample taken from a patient at diagnosis. Consistent amplification of *TEL–AML1* transcripts was observed at a dilution of $10^{-4}$, and occasionally at $10^{-5}$.

Assay sensitivity for *AML1–ETO* was estimated with 10-fold serial dilutions of a cDNA isolated from a Kasumi-1 cell line. *AML1-ETO* transcripts up to a $10^{-5}$ aqueous dilution were consistently detected.

When the sensitivity of the procedure was assessed using serial 10-fold dilutions of the plasmid pCR II-TOPO[AML1-ETO] (from $10^{-3}$ to $10^{-8}$), we established that the method can detect at least 0.05 fg plasmid, equivalent to 11 copies.

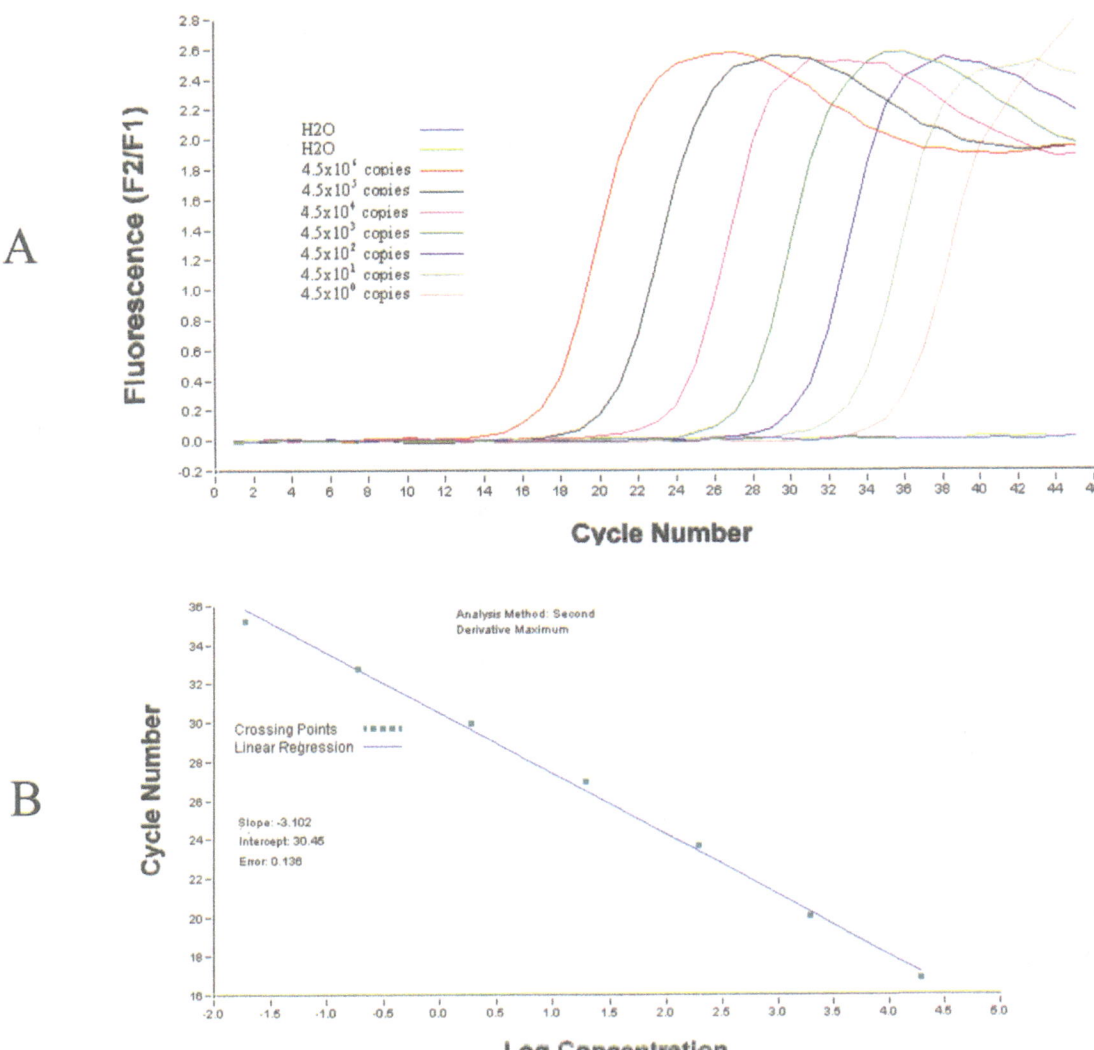

**Fig. 2. (A)** Standard curve for *TEL–AML1* using seven $10^{-1}$ serial dilutions of the plasmid pCR
II-TOPO$^{\text{TEL-AML1}}$. **(B)** Standard curve

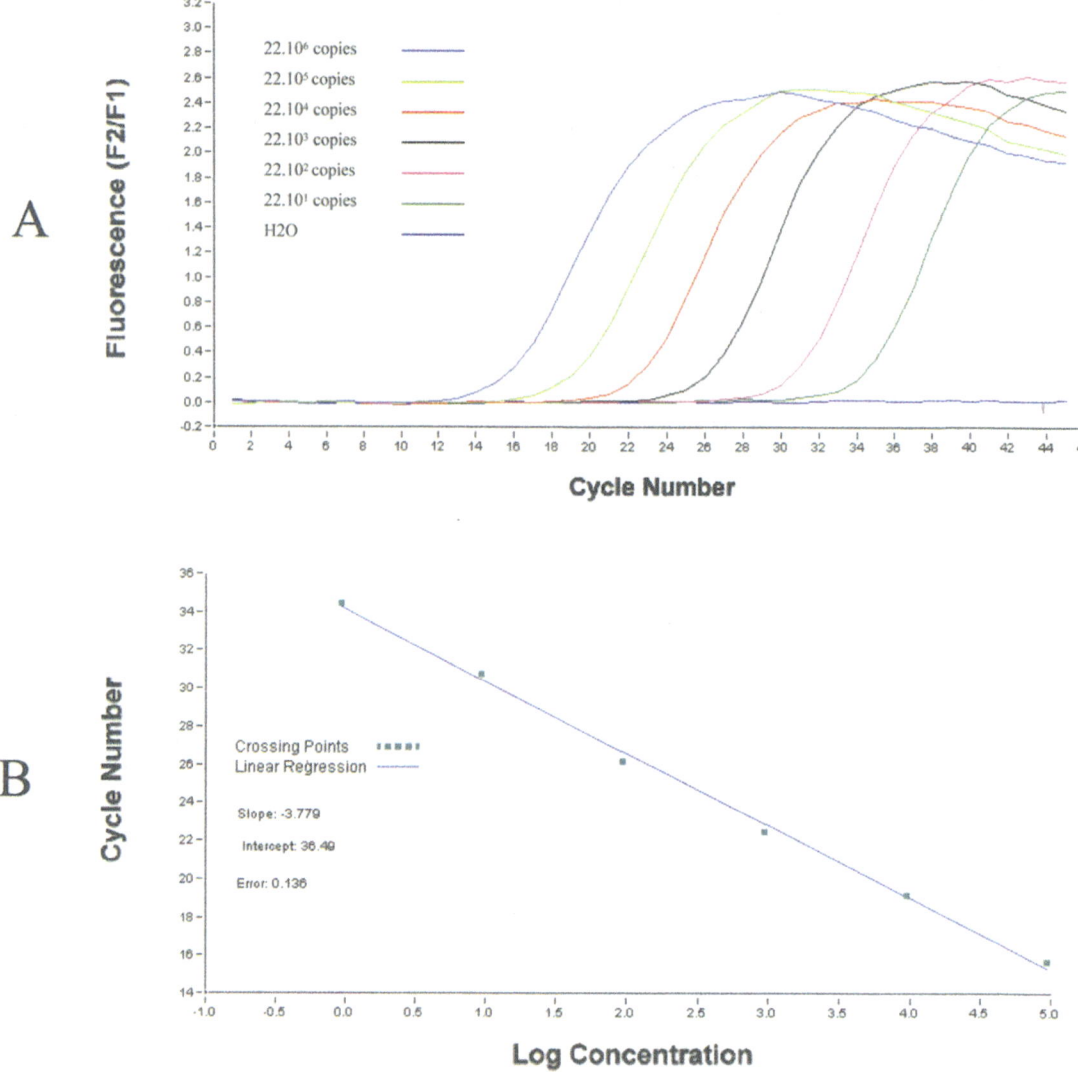

**Fig. 3. (A)** Standard curve for *AML1–ETO* using six 10⁻¹ serial dilutions of the plasmid pCR II-TOPO[AML1-ETO]. **(B)** Standard curve

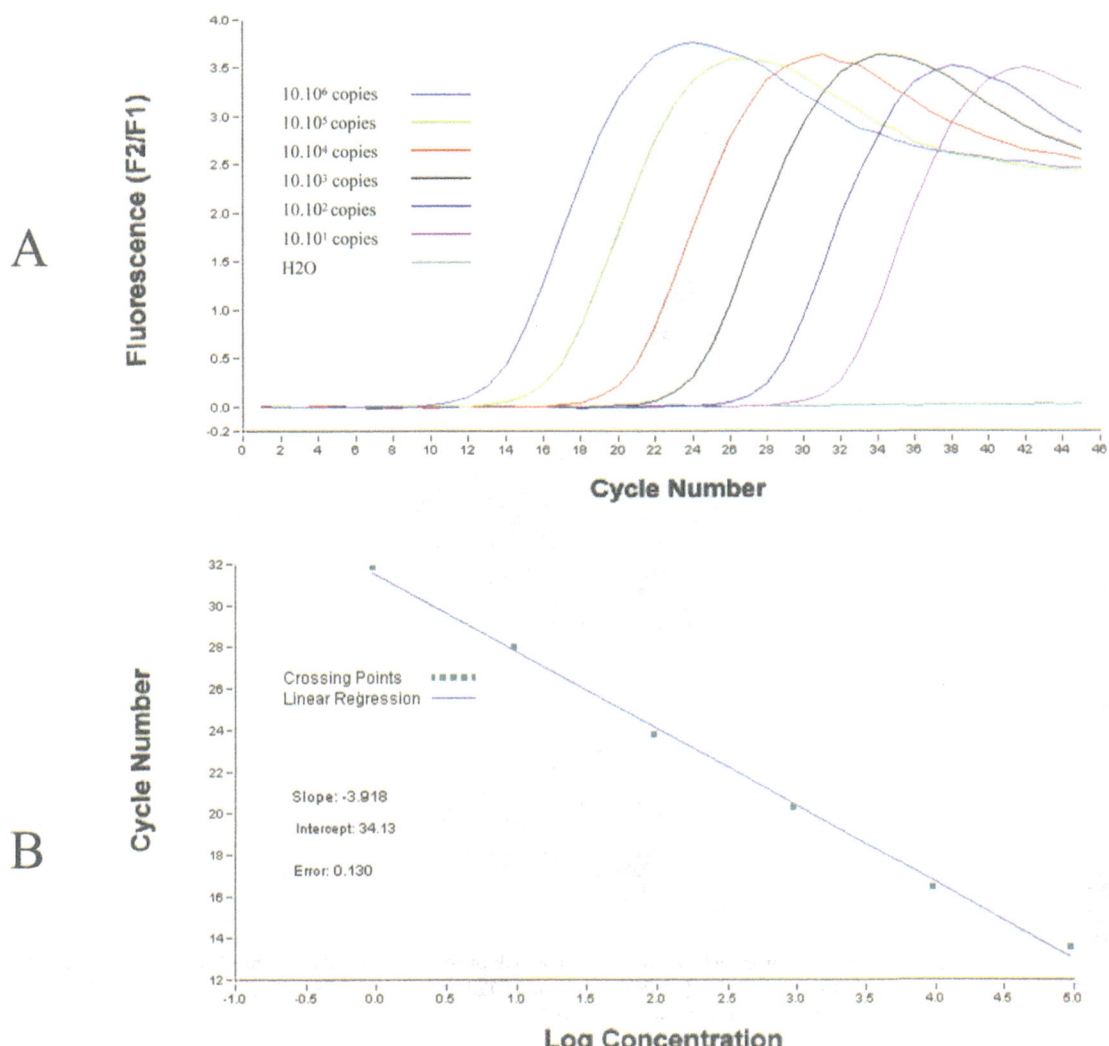

**Fig. 4.** (**A**) Standard curve for *AML1* using six $10^{-1}$ serial dilutions of the plasmid pCR II-TOPO^AML1. (**B**) Standard curve

**TEL-AML1**

A

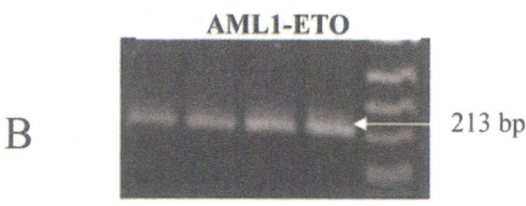

—134 bp

**AML1-ETO**

B

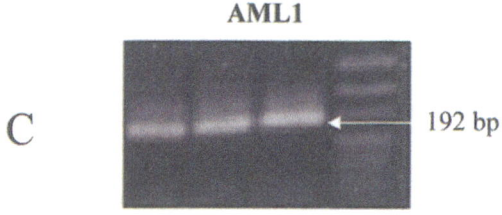

←213 bp

**AML1**

C

←192 bp

**Fig. 5.** Agarose gel electrophoresis of PCR products removed from the glass capillaries after PCR in the LightCycler. (**A**) PCR products of *TEL–AML1*. (**B**) PCR products of *AML1–ETO*. (**C**) PCR products of the control gene, *AML1*

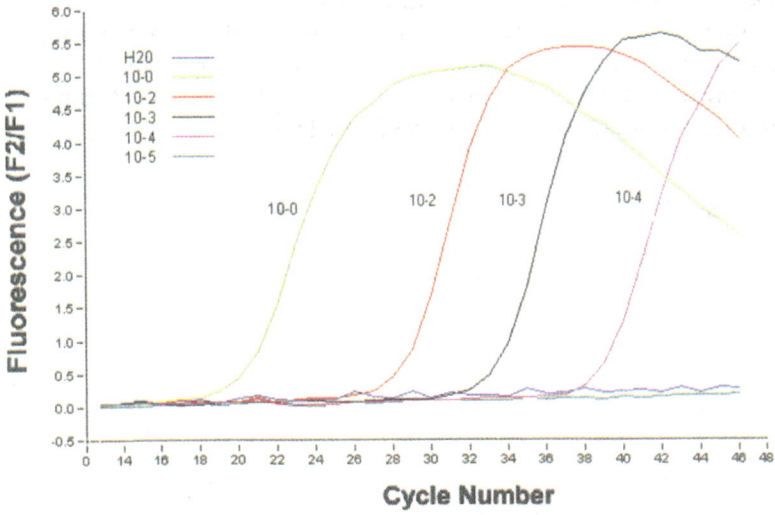

**Fig. 6.** *TEL–AML1* serial $10^{-1}$ dilutions of a positive cDNA sample from a patient at diagnosis. The fusion transcript was consistently detected at cDNA dilutions of $10^{-4}$

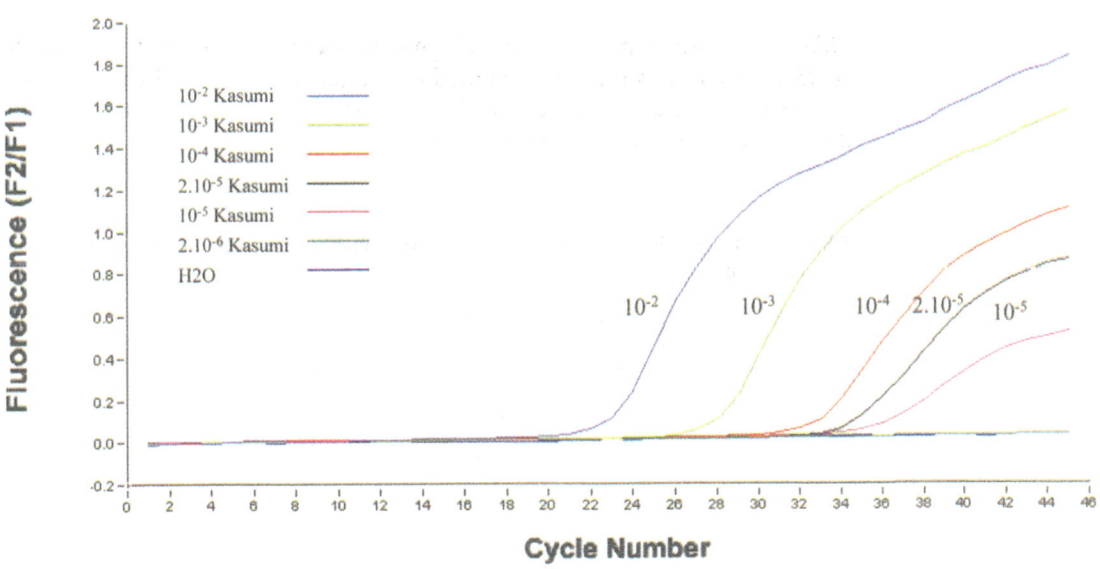

**Fig. 7.** *AML1–ETO* in serial $10^{-1}$ dilutions of a cDNA sample from Kasumi-1 cells. The fusion transcript was detected at cDNA dilutions of up to $10^{-5}$

**Reliability of the Assays**

Within-assay reproducibility for *TEL–AML1* was studied by repeating the analysis of the same sample 10 times in the same assay, calculating a coefficient of variance (cv) of 7% when the results were expressed as copies/μl cDNA. The between-assay reproducibility, estimated by analyzing the same cDNA sample in seven successive independent assays, showed a cv 13%.

**Table 4.** Assay reliability

| Assay | N | Cp mean±SD | copies/μl cDNA mean±SD(cv) |
|---|---|---|---|
| **TEL-AML1** | | | |
| Within-assay | 10 | 23.1±0.2 | $9.3 \times 10^5 \pm 6.8 \times 10^4$ (7) |
| Between-assay | 7 | 20.6±0.2 | $4.2 \times 10^6 \pm 5.3 \times 10^5$ (13) |
| **AML1-ETO** | | | |
| Within-assay | 10 | 21.43±0.07 | $1.2 \times 10^6 1 \pm 4.5 \times 10^4$ (4) |
| Between-assay | 10 | 20.31±0.24 | $2.1 \times 10^6 \pm 2.9 \times 10^5$ (14) |
| **AML1** | | | |
| Within-assay | 10 | 21.38±0.1 | $3.4 \times 10^5 \pm 2.3 \times 10^4$ (7) |
| Between-assay | 10 | 20.18±0.26 | $5.5 \times 10^5 \pm 6.6 \times 10^4$ (12) |

Within-assay reproducibility for *AML1–ETO* was assessed using the Kasumi-1 cell line at a $10^{-2}$ dilution. The within assay cv was 4% and the between assay cv, 14%.

The reference gene *AML1* had a within-assay cv of 7% and a between-assay cv of 12%.

**Quantitative Monitoring**

*TEL–AML1* transcripts from the 10 positive patients, expressed in copies/μl cDNA or as a ratio with *AML1,* showed a significant reduction of approximately four logs in the samples collected after treatment relative to those taken during the period of active disease, as seen in Table 5.

**Table 5.** Results of TEL-AML1 and AML1-ETO obtained in patients with active disease and after treatment

| Parameter | Active disease | After treatment | p |
|---|---|---|---|
| **TEL-AML1** | Median(range) | Median(range) | |
| TEL-AML1 (copies/μl cDNA) | $356 \times 10^3 (38 \times 10^{3-2} 15 \times 10^4)$ | $94 (0-2 \times 10^3)$ | 0.000 |
| TEL-AML1/AML1 (%) | 16.7(11.5–76.3) | 0.015(0.0–2.5) | 0.000 |
| **AML1-ETO** | | | |
| AML1-ETO (copies/μl cDNA) | $878 \times 10^3 (304 \times 10^3 - 192 \times 10^4)$ | $79 (0-4 \times 10^4)$ | 0.000 |
| $\Delta$ML1-ETO/$\Delta$ML1 (%) | 350(120–660) | 0.025(0–10%) | 0.000 |

In the quantitative studies carried out on eight *AML1–ETO* patients, the transcripts, expressed both as copies/µl cDNA or as a ratio with *AML1*, showed a log reduction greater than four in samples collected after treatment relative to those taken during the period of active disease.

## Comments

The first real-time PCR methods reported for the quantification of *TEL–AML1* [9,10] and *AML1–ETO* [21,22] were designed with TaqMan probes. However, the methods developed here are based on HybProbe technology adapted to the Light-Cycler equipment. Although TaqMan probes can be used in the LightCycler [23], HybProbes confer an advantage over TaqMan because they do not require long annealing–elongation incubation times (≥ 1 min) to achieve complete 5' nuclease degradation. This, combined with the high-speed thermal cycling characteristic of the LightCycler, allows very rapid results.

The *TEL–AML1*, *AML1–ETO*, and *AML1* standard curves were highly reproducible in successive assays, yielding very consistent $C_p$ values for each concentration, and very high coefficients of correlation.

The sensitivity of *TEL–AML1* detection is similar to that achieved by other methods using TaqMan probes [10]. The sensitivity of *AML1–ETO* detection was one log higher than that for *TEL–AML1*, also equivalent to other qualitative methods [20] and TaqMan quantitative methods [21,22]. The levels of sensitivity reported here for *TEL–AML1* and *AML1–ETO* are satisfactory for monitoring MRD and for verifying the effectiveness of therapeutic regimes.

The between-assay cv of 13–14% reported here for *TEL–AML1*, *AML1–ETO*, and *AML1* is similar to those reported for equivalent methods on other systems [10,21,22]. This precision is sufficiently reliable to monitor sequential samples from each patient.

*TEL–AML1* and *AML1–ETO* transcripts decreased by at least four logs after initial treatment. Nevertheless, transcripts were still detected in some samples collected soon after induction. The levels of transcripts detected at this time were very low, and the vast majority of transcripts had disappeared by follow-up. However, a large increase in the number of *AML–ETO* transcripts preceded relapse in those patients who suffered relapse. The present data confirms the usefulness of the quantitative assessment of *TEL–AML1* and *AML1–ETO* transcripts to monitor the efficacy of a patient's response to therapy and in MRD monitoring.

## Applications

Our results confirm that real-time PCR methods using the LightCycler and HybProbes to quantify *TEL–AML1* and *AML1–ETO* transcripts, are reliable for the diagnosis and monitoring of MRD in leukemia patients carrying t(12;21) or t(8;21) translocations.

## Acknowledgements

This study has been possible with the assistance of Spanish FIS grant 99/0806.

# References

1. Nucifora G, Rowley J (1995) AML1 and the 8;21 and 3;21 translocations in acute and chronic myeloid leukemia. Blood 86: 1–14
2. Grimwade D, Walker H, Oliver F, Wheatly K, Harrison C, Harrison G, Rees J, Hann I, Stevens R, Burnett A, Goldstone A. (1998) The importance of diagnostic cytogenetics on outcome in AML: analysis of 1612 patients entered into the MRC AML 10 Trial. Blood 92: 2322–2333
3. Mauvieux L, Delabesse E, Macintyre E (1997) Molecular genetics of acute leukemias. Rev Clin Exp Hematol 2: 3–26
4. Camitta BM, Pullen J, Murphy S (1997) Biology and treatment of acute lymphoblastic leukemia in children. Semin Oncol 24: 83–91
5. McLean TW, Ringold S, Neuberg D, Stegmaier K, Tantravahi R, Ritz J, Koeffler HP, Takeuchi S, Janssen JW, Seriu T, Bartram CR, Sallan SE, Gilliland DG, Golub TR. (1996) TEL/AML1 dimerizes and is associated with a favorable outcome in childhood acute lymphoblastic leukemia. Blood 88: 4252–4258
6. Morschhauser F, Cayuela JM, Martini S, Baruchel A, Pousselot P, Socié G, Berthou P, Jouet JP, Straetmans, Sigaux F, Fenaux P, Preudhome C (2000) Evaluation of minimal residual disease using reverse-transcription polymerase chain reaction in t(8;21) acute myeloid leukemia: a multicenter study of 51 patients. J Clin Oncol 18: 788–794
7. Marcucci G, Livak KJ, Bi W, Strout MP, Bloomfield CD, Caligiuri MA (1998) Detection of minimal residual disease in patients with AML1/ETO-associated acute myeloid leukemia using a novel quantitative reverse transcription polymerase chain reaction. Leukemia 12: 1482–1489
8. Wattjes MP, Krauter J, Nagel S, Heidenreich O, Ganser A, Heil G (2000) Comparison of nested competitive RT–PCR and real-time RT–PCR for the detection and quantification of AML1/MTG8 fusion transcripts in t(8;21) positive acute myelogenous leukemia. Leukemia 14: 329–335
9. Pallisgaard N, Clausen N, Schroder H, Hokland P (1999) Rapid and sensitive minimal residual disease detection in acute leukemia by quantitative real-time RT–PCR exemplified by t(12;21) *TEL-AML1* fusion. Genes Chromosomes Cancer 26: 355–365
10. Ballerini P, Landman Parker J, Laurendeau I, Olivi M, Vidaud M, Adam M, Leverger G, Gerota I, Cayre YE, Bieche I (2000) Quantitative analysis of TEL/AML1 fusion transcripts by real-time RT–PCR assay in childhood acute lymphoblastic leukemia. Leukemia 14: 1526–1531
11. Bolufer P, Barragán E, Verdeguer A, Cervera J, Fernández JM, Moreno I, Lerma E, Esquembre C, Tasso M, Fuster V, Bermúdez M, Sanz MA (2002) Rapid quantitative detection of TEL-AML1 fusion transcripts in acute lymphoblastic leukemia by real-time reverse transcription PCR using fluorescently labeled probes. Haematologica 87: 32–41
12. Barragan E, Bolufer P, Moreno I, Martin G, Nomdedeu J, Brunet S, Fernandez P, Rivas C, Sanz MA. (2001) Quantitative detection of AML1-ETO rearrangement by real-time RT–PCR using fluorescently labeled probes. Leukemia & Lymphoma 42: 747–756
13. Chomczynski P, Sacchi N (1987) Single-step method of RNA isolation by acid guanidinium thiocyanate–phenol–chloroform extraction. Anal Biochem 162: 156–159
14. Satake N, Kobayashi H, Tsunematsu Y, Kawasaki H, Koizumi S, Kaneko Y (1997) Minimal residual disease with TEL-AML1 fusion transcript in childhood acute lymphoblastic leukaemia with t(12;21). Br J Haematol 97: 607–611
15. Satake N, Kobayashi H, Tsunematsu Y, Kawasaki H, Koizumi S, Kaneko Y (1993) Junctions of the AML1/MTG8(ETO) fusion are constant in t(8;21) acute myeloid leukemia detected by reverse transcription polymerase chain reaction. Blood 82: 1270–1276
16. Nakao M, Yokota S, Horiike S, Kashima K, Sonoda Y, Fujimoto T, Misawa S (1996) Detection and quantification of TEL/AML1 fusion transcripts by polymerase chain reaction in childhood acute lymphoblastic leukemia. Leukemia 10: 1463–1470

17. van Dongen JJM, Macintyre EA, Gabert JA, Delabesse E, Rossi V, Saglio G, Gottardi E, Ramboldi A, Dotti G, Griesinger F, Parreira A, Gameiro P, Diaz MG, Molec M, Langerak AW, San Miguel JF, Biondi A (1999) Standardized RT–PCR analysis of fusion gene transcripts from chromosome aberrations in acute leukemia for detection of minimal residual disease. Leukemia; 13: 1901–1928

18. Nakao M, Yokota S, Horiike S, Kashima K, Sonoda Y, Fujimito T, Misawa S (1997) Detection and quantification of TEL/AML1 fusion transcripts by polymerase chain reaction in childhood acute lymphoblastic leukemia. Leukemia 10: 1463–1470

19. van Dongen JJM, Macintyre EA, Gabert JA, Delabesse E, Rossi V, Saglio G, Gottardi E, Ramboldi A, Dotti G, Griesinger F, Parreira A, Gameiro P, Díaz MG, Molec M, Lagerak AW, San Miguel JF, Biondi A (1999) Standardized RT–PCR analysis of fusion transcripts from chromosome aberrations in acute leukemia for detection of minimal residual disease. Leukemia 13: 1901–1928

20. Satake N, Kobayashi H, Tsunematsu Y, Kawasaki H, Horikoshi Y, Koizumi S, Kaneko Y (1995) Disappearance of AML1-MTG8 (ETO) fusion transcript in acute myeloid leukaemia patients with t(8;21) in long-term remission. Br J Haematol 91: 892–898

21. Marcucci G, Livak KJ, Bi W, Strout MP, Bloomfield CD, Caligiuri MA (1998) Detection of minimal residual disease in patients with AML1/ETO-associated acute myeloid leukemia using a novel quantitative reverse transcription polymerase chain reaction. Leukemia 12: 1482–1489

22. Wattjes MP, Krauter J, Nagel S, Heidenreich O, Ganser A, Heil G (2000) Comparison of nested competitive RT-PCR and real-time RT-PCR for the detection and quantification of AML1/MTG8 fusion transcripts in t(8;21) positive acute myelogenous leukemia. Leukemia 14: 329–335

23. Kreuzer KA, Lass U, Bohn A, Landt O, Schmidt CA (1999) LightCycler technology for the quantitation of bcr/abl fusion transcripts. Cancer Res 59: 3171–3174

# Applications

**II**

## Genetics

# Rapid Detection of Gene Duplications in Charcot-Marie-Tooth 1A Disease by SNP Genotyping Using Real-Time PCR

C. Ruiz-Ponte, A. Vega, L. Loidi, A. Carracedo, F. Barros*

## Introduction

Many clinical disorders arise from insertions, deletions, translocations, amplifications, or expansions in the genome. An increase in gene dosage may activate proto-oncogenes in cancer cells and can cause inherited diseases. Chromosomal aneuploidies demonstrate the consequences incorrect gene dosage.

Current methods for genetic quantification capable of discriminating between one, two, and more gene copies in human cells are generally expensive, time consuming, and require highly trained personnel for their performance and interpretation. Moreover, large amounts of high quality material are necessary for their completion.

Southern blots and pulse-field gel electrophoresis are widely used in clinical diagnostics to assess gene dosage for evaluation of chromosomal submicroscopic deletions and duplications. Fluorescent in situ hybridization assay (FISH) is available in only a limited number of clinical laboratories. More recent methods include comparative genomic hybridization against arrayed genomic clones [1], determination of relative target amounts by quantitative PCR using real-time analysis of logarithmic-phase amplicon accumulation [2], and a non-PCR-based light-activated interstrand nucleic acid cross-linking system [3].

We have used the deletion / duplication of a specific 1.4 Mb region at chromosome 17p11.2-p12 as a model system for developing techniques for gene dosage [4]. This region has a tendency for spontaneous deletion and duplication events of a characteristic size due to the presence of flanking low-copy repetitive regions. The autosomal dominant demyelinating peripheral neuropathy Charcot-Marie-Tooth disease type 1A (CMT1A) is characterized by slow progressive weakness and atrophy of distal muscles in the feet and/or hands, often associated with depressed tendon reflexes and loss of sensation. Nerve conduction velocities are almost always slow. The features of hereditary neuropathy with liability to pressure palsies (HNPP) comprise a disorder of peripheral nerves in which individuals are predisposed to repeated pressure neuropathies, such as carpal tunnel syndrome, and peroneal palsy with foot drop.

* F. Barros, Unidad de Medicina Molecular, Edificio de consultas planta -2, Hospital Clínico Universitario, 15706 Santiago de Compostela, Spain, E-mail: apimlbar@usc.es

In both syndromes, 80% of affected individuals carry duplications or deletions of chromosome 17p11.2-p12. CMT1A is associated with a tandem DNA duplication while HNPP is associated with a deletion of the same region [5, 6]. The peripheral myelin protein 22 (*PMP22*) gene is located within the 1.4 Mb CMT1A monomer. Although mutations within the *PMP22* gene have been identified in some cases of CMT1A and HNPP, a gene dosage effect has been proposed as the principal underlying mechanism for development of CMT1A and HNPP [7].

Molecular diagnosis of CMT1A involves detection of DNA duplication located within 17p11.2-p12. Southern blots and pulsed-field gel electrophoresis are the methods currently used to assess gene dosage [8]. Other methods include FISH [9], and analysis of polymorphic markers located within the region [10].

We previously described a strategy to determine gene dosage in CMT1A and HNPP patients based on competitive real-time fluorescent PCR and single nucleotide polymorphism (SNP) genotyping by analyzing their melting temperature [4]. In CMT1A samples, when heterozygous for a SNP, the ratio between the areas under the melting curves determines relative concentrations for both alleles and reveals whether the individual has duplication. Furthermore, a heterozygous status rules out the HNPP condition.

A single SNP is not sufficient because only heterozygous individuals are informative for the duplication assay. In the best situation (i.e., an SNP with 50% heterozygosity) one SNP will only be informative in 50% of affected patients. More SNPs are necessary to increase the chance of definitively determining the gene dosage.

## Materials

Equipment
LightCycler® instrument (Roche Diagnostics, Mannheim, Germany)
LightCycler® software, version 3.3 (Roche Diagnostics, Mannheim, Germany)

Reagents
Amplification primers (TIB MOLBIOL, Berlin, Germany)
Hybridization probes (TIB MOLBIOL, Berlin, Germany)
LightCycler DNA Master Hybridization Probes (Roche Diagnostics, Mannheim, Germany)
Wizard® Genomic DNA Purification Kit (Promega, Madison, USA)

## Procedure

DNA Template Preparation
Genomic DNA was isolated by a standard rapid lysis technique using the Wizard Genomic DNA Purification Kit. A total of 100 normal samples and 21 positive control samples for duplication were analyzed.

SNP Search and Selection
Ruiz-Ponte et al [4] developed a gene dosage approach for detection of *PMP22* gene duplications using the SNP PMPIVS3+33. We searched for additional SNPs, both within, and surrounding the *PMP22* gene. The search was performed using

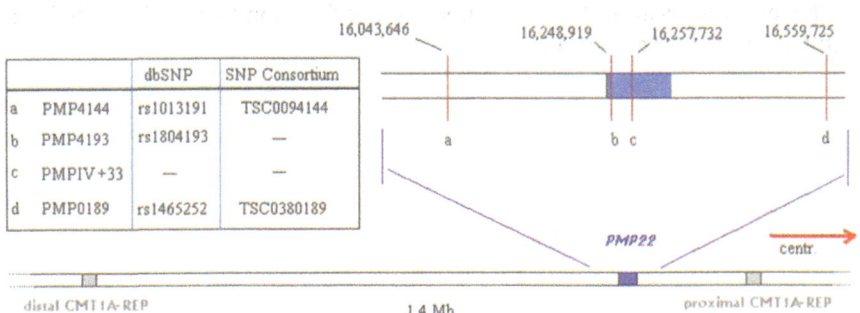

**Fig. 1.** Scheme of the 17p11.2 region with the positions of the *PMP22* gene and the SNPs indicated (refered to SNP Consortium database)

the two main SNP repositories: The SNP consortium (http://snp.cshl.org/) and dbSNP (http://www.ncbi.nlm.nih.gov/SNP/). Although more than 71 variations in this region are already known and described in these databases, only a few are validated with known allele frequencies. We chose two validated SNPs near the PMP22 gene (rs1013191 and rs1465252). In addition, the polymorphism rs1804193 was validated in our lab using a panel of 50 healthy control samples. SNPs used in this work and their positions are listed in Figure 1.

Primers and probes used for genotyping the SNP PMPIVS3+33 were previously published [4]. All hybridization probes were designed by TIB MOLBIOL. Details are shown in Table 1.

**Primer and Probe Design**

The following master mixes were used for amplification and hybridization based detection of the SNPs:

**LightCycler PCR**

| | Volume [µl] | [Final] |
|---|---|---|
| LightCycler DNA Master Hybridization Probes | 2 | 1 X |
| MgCl$_2$ stock solution (25 mM) | 2.4 | 4 mM |
| Primers (10 µM each) | 0.7 + 0.7 | 0.4 µM |
| Hybridization probes (4 µM each) | 0.5 + 0.5 | 0.1 µM |
| H$_2$O (PCR grade) | 11.2 | |
| Total volume | 18 | |

The amplification mixture was completed with 18 µl of the master mix and 2 µl of the corresponding DNA template preparation, and was added to each capillary. After a pulse centrifugation in a microcentrifuge to fill the cuvettes, sealed capillaries were placed into the LightCycler rotor. In each set of experiments, a normal heterozygous control and duplicate heterozygous control were included. The negative control sample was prepared by adding PCR-grade water instead of the DNA template.

The following PCR protocol was used for amplification (for clarity, individual LightCycler settings are given):

Experimental protocol for PMPIVS3 +33 SNP
- Denaturation for 2 min at 95°C
- Amplification

| Parameter | Value | | |
|---|---|---|---|
| Cycles | 31 | | |
| Type | None | | |
| | Segment 1 | Segment 2 | Segment 3 |
| Target temperature [°C] | 95 | 55 | 72 |
| Incubation time [s] | 0 | 10 | 5 |
| Temperature transition rate [°C/s] | 20 | 20 | 20 |
| Acquisition mode | None | Single | None |
| Fluorescence Display Mode | F2 / 1 | | |

- Melting Curve Analysis

| Parameter | Value | | |
|---|---|---|---|
| Cycles | 1 | | |
| Type | Melting | | |
| | Segment 1 | Segment 2 | Segment 3 |
| Target temperature [°C] | 95 | 55 | 80 |
| Incubation time [s] | 5 | 10 | 0 |
| Temperature transition rate [°C/s] | 20 | 20 | 0.2 |
| Acquisition mode | None | None | Continue |
| Fluorescence Display Mode | F2 / 1 | | |

- Cooling for 2 min at 40°C.

Experimental protocol for PMP0189 SNP
- Denaturation for 2 min at 95°C
- Amplification

| Parameter | Value | | |
|---|---|---|---|
| Cycles | 40 | | |
| Type | None | | |
| | Segment 1 | Segment 2 | Segment 3 |
| Target temperature [°C] | 95 | 52 | 72 |
| Incubation time [s] | 0 | 5 | 5 |
| Temperature transition rate [°C/s] | 20 | 20 | 20 |
| Acquisition mode | None | Single | None |
| Fluorescence Display Mode | F2 / 1 | | |

- Melting Curve Analysis

| Parameter | Value | | |
|---|---|---|---|
| Cycles | 1 | | |
| Type | Melting | | |
| | Segment 1 | Segment 2 | Segment 3 |
| Target temperature [°C] | 95 | 37 | 75 |
| Incubation time [s] | 5 | 10 | 0 |
| Temperature transition rate [°C/s] | 20 | 20 | 0.1 |
| Acquisition mode | None | None | Continuous |

- Cooling for 2 min at 40°C.

Experimental protocol for PMP4144 SNP
- Denaturation for 2 min at 95°C
- Amplification

| Parameter | Value | | |
|---|---|---|---|
| Cycles | 40 | | |
| Type | None | | |
| | Segment 1 | Segment 2 | Segment 3 |
| Target temperature [°C] | 95 | 50 | 72 |
| Incubation time [s] | 0 | 5 | 5 |
| Temperature transition rate [°C/s] | 20 | 20 | 20 |
| Acquisition mode | None | Single | None |
| Fluorescence Display Mode | F2 / 1 | | |

- Melting Curve Analysis

| Parameter | Value | | |
|---|---|---|---|
| Cycles | 1 | | |
| Type | Melting | | |
| | Segment 1 | Segment 2 | Segment 3 |
| Target temperature [°C] | 95 | 37 | 85 |
| Incubation time [s] | 5 | 10 | 0 |
| Temperature transition rate [°C/s] | 20 | 20 | 0.1 |
| Acquisition mode | None | None | Continuous |

- Cooling for 2 min at 40°C.

Experimental protocol for PMP4193 SNP
- Denaturation for 2 min at 95°C
- Amplification

| Parameter | Value | | |
|---|---|---|---|
| Cycles | 31 | | |
| Type | None | | |
| | Segment 1 | Segment 2 | Segment 3 |
| Target temperature [°C] | 95 | 55 | 72 |
| Incubation time [s] | 0 | 10 | 5 |
| Temperature transition rate [°C/s] | 20 | 20 | 20 |
| Acquisition mode | None | Single | None |
| Fluorescence Display Mode | F2 / 1 | | |

- Melting Curve Analysis

| Parameter | Value | | |
|---|---|---|---|
| Cycles | 1 | | |
| Type | Melting | | |
| | Segment 1 | Segment 2 | Segment 3 |
| Target temperature [°C] | 95 | 42 | 80 |
| Incubation time [s] | 5 | 10 | 0 |
| Temperature transition rate [°C/s] | 20 | 20 | 0.2 |
| Acquisition mode | None | None | Continuous |

- Cooling for 2 min at 40°C.

**Melting Curve Areas Analysis**

Melting curve data were analyzed with LightCycler software by plotting the rate of fluorescence change (-dF/dT) *vs.* temperature (T). The areas under the melting curves were calculated using LightCycler software. The ratio of the areas under the melting curves for the heterozygous normal control was normalized to a value of 1.0, and the samples were scaled accordingly. All calculations were made using the following formula:

$$Rs = As * Bc / Ac * Bs$$

where Rs is the normalized ratio in the sample, As is the area under the peak in the sample that may be duplicated, Ac the area under the same peak in the control, and Bs and Bc are the areas under the reference peaks in the sample and control respectively.

**Table 1.** Oligonucleotides

| PMPIVS3+33 SNP (GenBank Accesion #AC005703) | | | | |
|---|---|---|---|---|
| | Position | Length | GC (%) | $T_m$ (°C) |
| Primers | | | | |
| CCATGGCCAGCTCTCCTAAC | 186862–186881 | 20 | 60 | 66.16 |
| CATTCCGCAGACTTTGATGC | 187099–187080 | 20 | 50 | 63.58 |
| Product | | 192 | | |
| Probes | | | | |
| TTCCAAATTCTTGCTGGTAAGT-TGTGGAT -F | 187027–187055 | 29 | 37.9 | 68.52 |
| LCRed640-TAAAGTCCATGTGGA-AGCGGGGT | 187058–187080 | 23 | 52.2 | 69.70 |

| PMP 4193 SNP (GenBank Accesion #NM_153322) | | | | |
|---|---|---|---|---|
| Primers | | | | |
| GCTGTTGATTGAAGATGTAT | 675–694 | 20 | 35 | 57.05 |
| GGATGTAAAGTTCCTTAGC | 824–806 | 19 | 42.1 | 56.48 |
| Product | | 150 | | |
| Probes | | | | |
| ATGTACATAGTATTGTTTACTTT-TTATG-F | 742–769 | 28 | 21.4 | 58.06 |
| LCRed640-TGACCATCAGCCTCG-TGTTGAGCCT | 771–795 | 25 | 56.0 | 73.31 |

| PMP0189 (GenBank Accesion #AC005517) | | | | |
|---|---|---|---|---|
| Primers | | | | |
| TTTGGGATTAAATAATGTCCAGG | 29123–29145 | 23 | 34.8 | 60.09 |
| AAAGTGGGAAGGTACAGGATGTA | 29309–29287 | 23 | 43.5 | 64.42 |
| Products | | 185 | | |
| CCCTTCCCGTTGACTCAT-F | 29167–29185 | 18 | 55.6 | 62.66 |
| LCRed640-TCAACTTGGGATTCAC-TATGACCACTAG | 29187–29214 | 28 | 42.9 | 67.27 |

| PMP4144 (GenBank Accesion #AC005772) | | | | |
|---|---|---|---|---|
| Primers | | | | |
| TCAGACAAGTTGGGTCAAACAA | 5941–5920 | 22 | 40.9 | 63.95 |
| GGGACTCCTACCGTTCAGTGT | 5634–5654 | 21 | 57.1 | 67.00 |
| Products | | 306 | | |
| Probes | | | | |
| ATGGTTTAACTAAGTAGCAAGA-AAAGGAC-F | 5742–5714 | 29 | 34.5 | 65.04 |
| LCRed640-GGTTCTCAGTTAATT-ATTAAATTCTCAAA | 5712–5684 | 29 | 24.1 | 60.30 |

## Results

We established PCR protocols for four SNPs in the 17p11.2-p12 region. Figures 2–5 show the melting curve analysis of each polymorphism. Genotype assignments were verified by sequencing on a MegaBACE ™ 1000 sequencer (Amersham Biosciences, Uppsala, Sweden).

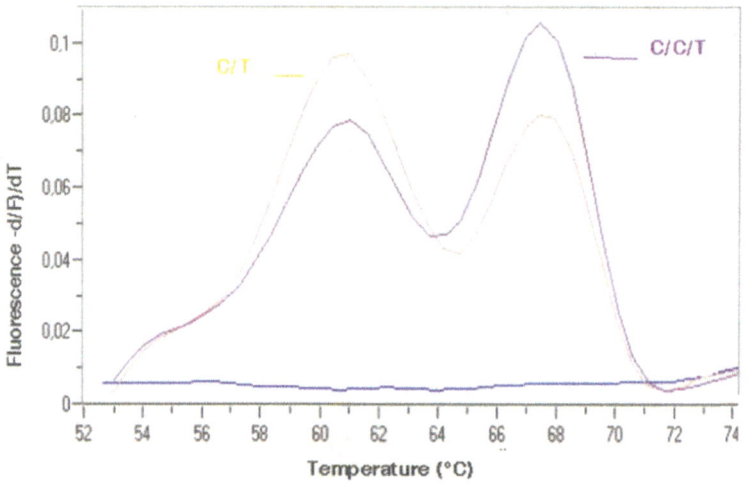

**Fig. 2.** Melting curves of two heterozygous genotypes of the PMPIVS3+33 SNP. The sensor probe is complementary to the C allele ($T_m$=67°C). The T allele has a $T_m$ of 60°C

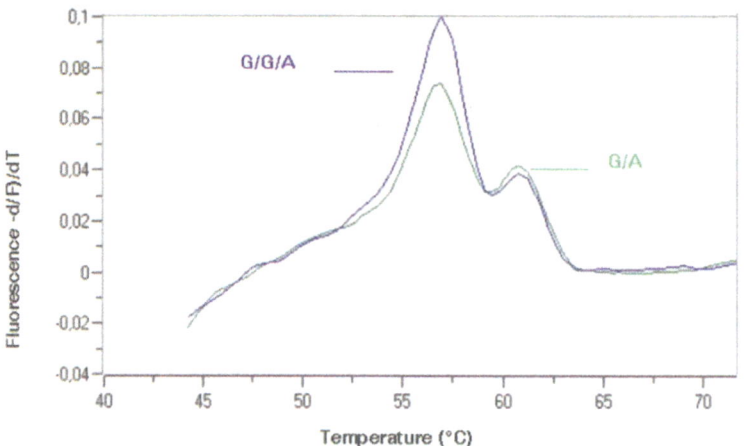

**Fig. 3.** Melting curve analysis of two heterozygous genotypes of the rs1804193 SNP. Sensor probe is complementary to the A allele ($T_m$=61°C). The G allele has a $T_m$ of 57.6°C

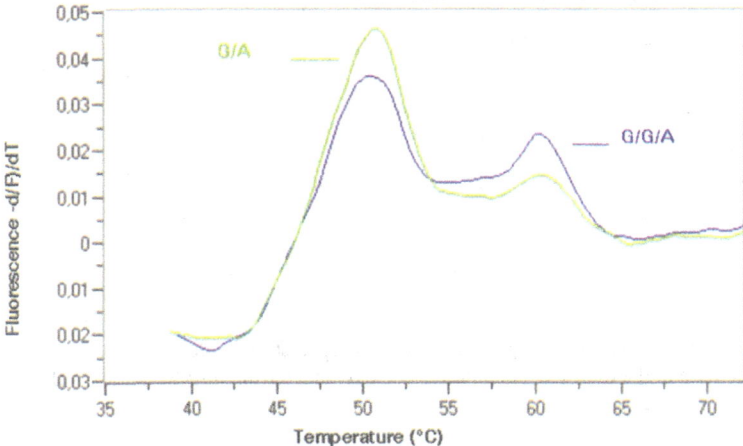

**Fig. 4.** Melting curves of two heterozygous genotypes of the rs1465252 SNP. The sensor probe is complementary to the G allele ($T_m$=62°C). The A allele has a $T_m$ of 52°C

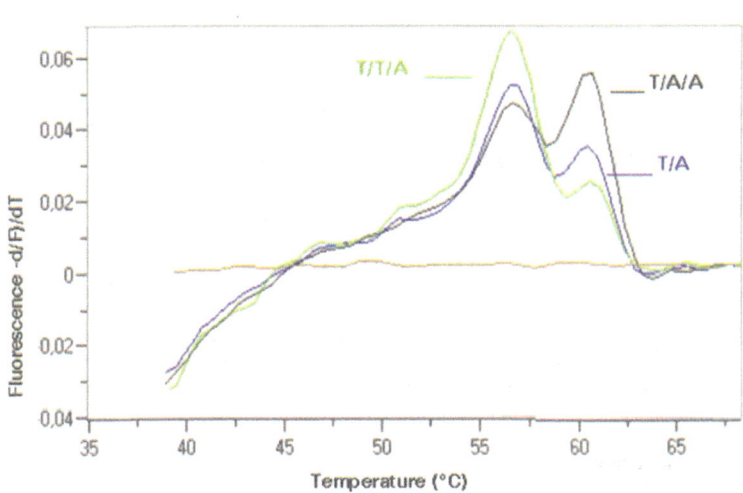

**Fig. 5.** Melting curves for the rs1013191 SNP showing three heterozygous genotypes. The sensor probe is complementary to the A allele ($T_m$=60°C). The T allele has a $T_m$ of 57°C

A summary for each SNP with their $T_m$s and frequencies is shown in Table 2.

Figures 2–5 show results with normal heterozygous samples and duplicated samples. We were able to detect duplications in all positive sample controls. The mean of ratios between the areas under the melting curves are reported in Table 3.

**Table 2.** Observed melting temperatures and allele frequencies

| SNPs | Variant alleles | Allele | $T_m$ (°C) | Frequence |
|------|-----------------|--------|------------|-----------|
| PMPIVS3+33[a] | C/T | T | 60 | 0.17 |
| | | C | 67 | 0.83 |
| PMP4193[a] | G/A | G | 57.6 | 0.91 |
| | | A | 61 | 0.09 |
| PMP0189[b] | G/A | A | 52 | 0.5 |
| | | G | 62 | 0.5 |
| PMP4144[b] | A/T | T | 57 | 0.52 |
| | | A | 60 | 0.48 |

[a] Allelic frequencies determined on a panel of 50 healthy Spanish individuals
[b] Allelic frequencies in the Caucasian population (data obtained from http://snp.cshl.org)

**Table 3.** Ratios (mean ± standard deviation) of the areas under the melting curves of normal and duplicated samples

| SNPs | Sample | Area ratios |
|------|--------|-------------|
| PMP IVS3+33 | Normal n=34 | 0.96 ± 0.01 |
| | Duplicated n=17 | 1.45 ± 0.05 |
| PMP4193 | Normal n=13 | 1.02 ± 0.13 |
| | Duplicated n=7 | 1.75 ± 0.18 |
| PMP0189 | Normal n=13 | 1.12 ± 0.07 |
| | Duplicated n=10 | 1.70 ± 0.05 |
| PMP4144 | Normal n=17 | 1.07 ± 0.17 |
| | Duplicated n=13 | 1.77 ± 0.28 |

n = number of samples analysed

## Comments

Despite the 1:1 relationship of alleles in normal heterozygous samples, we observed a ratio between the areas under the melting curves that often differed from 1. This is due to small changes in PCR conditions, such as probe concentration from one experiment to another. This potential problem is easily solved by including a normal heterozygous control in each genotyping assay. As we discussed elsewhere [4], melting curve area ratios of the same sample showed a greater variation between different PCR assays than among replicates within the same PCR. Specifically, in the same PCR assay, the ratio of the sample area must be normalized to the ratio of a normal heterozygous control. Following our proposed protocol, accurate calculation of ratios is not possible without a normal heterozygous control.

For the SNP PMP4144 we obtained three possible heterozygous genotypes: A/T, A/A/T, and A/T/T. As shown in Figure 4, the melting curves are clearly distinct for each case. However, when the ratio of the areas under the melting curves was calculated for the T/A/A genotype, the value obtained was similar to the normal heterozygous control. In this case, using the peak height instead of the area would provide more accurate results.

SNPs genotyping by real-time fluorescent PCR is an easy, rapid, and standardized method that can offer accurate gene dosage determination in heterozygous individuals. For samples suspected of carrying a deletion, heterozygous genotypes indicate the absence of deletion.

By using the four SNPs described, the probability of finding one heterozygous locus is approximately 85%. To assure rapid diagnosis of nearly all patients, a panel of SNPs with a 98% probability of finding two informative SNPs would be ideal.

## References

1. Snijders AM, Nowak N, Segraves R, Blackwood S, Brown N, Conroy J, et al. (2001) Assembly of microarrays for genome-wide measurement of DNA copy number. Nat Genet 29:263–264
2. Feldkotter M, Schwarzer V, Wirth R, Wienker TF, Wirth B (2002) Quantitative analyses of SMN1 and SMN2 based on real-time LightCycler PCR: fast and highly reliable carrier testing and prediction of severity of spinal muscular atrophy. Am J Hum Genet 70:358–368
3. Peoples R, Weltman H, Van Atta R, Wang J, Wood M, Raimondi MF, Cheng P, Huan B (2002) High-Throughput Detection of Submicroscopic Deletions and Methylation Status at 15q11-q13 by a Photo-Cross-Linking Oligonucleotide Hybridization Assay. Clin Chem 48:1844–1850
4. Ruiz-Ponte C, Loidi L, Vega A, Carracedo A, Barros F (2000) Rapid Real-Time Fluorescent PCR Gene Dosage Test for the diagnosis of DNA duplications and deletions. Clin Chem 46:1574–1582
5. Lupski JR, de Oca-Luna RM, Slaugenhaupt S, Pentao L, Killian JM, Garcia CA, et al. (1991) DNA duplication associated with Charcot-Marie-Tooth disease type 1A. Cell 66:219–232
6. Chance PF, Alderson MK, Leppig KA, Lensch MW, Matsunami N, Smith B, et al. (1993) DNA deletion associated with hereditary neuropathy with liability to pressure palsies. Cell 72:143–151
7. Lupski JR, Wise CA, Kuwano A, Pentao L, Parker JT, Glaze DG, et al. (1992) Gene dosage is a mechanism for Charcot-Marie-Tooth disease type 1A. Nat Genet 1:29–33
8. Lupski JR (1996) DNA diagnosis for Charcot-Marie-Tooth disease and related inherited neuropathies. Clin Chem 42:995–998
9. Shaffer LG, Kennedy GM, Spliker AS, Lupski JR (1997) Diagnosis of the CMT1A duplication and HNPP deletion by interphase FISH: impli-cations for testing in the cytogenetic laboratory. Am J Med Genet 69:325–331
10. Badano JL, Inoue K, Katsanis N, Lupski JR (2001) New polymorphic short tandem repeats for PCR-based Charcot-Marie-Tooth disease type 1A duplication diagnosis. Clin. Chem 47:838–843

# Cytochrome P450 2D6 Deletion Genotyping Using Derivative Curve Analysis on the LightCycler

Alison Millson, Elizabeth L. Frank, Elaine Lyon*

## Introduction

Cytochrome P450 2D6 (CYP2D6) is an isoenzyme of the cytochrome P450 super family that catalyzes the oxidative biotransformation of numerous exogenous compounds. These compounds include many commonly prescribed drugs such as antiarrhythmics, β-receptor blockers, neuroleptics, selective serotonin reuptake inhibitors and tricyclic antidepressants (1). Genetic polymorphism in CYP2D6 is common and can affect therapeutic response. Polymorphisms include single nucleotide polymorphisms (SNPs), gene deletions and duplications. When gene deletions or SNPs are present, drug clearance is impaired, resulting in possible accumulation and toxicity of the drug. When the gene is duplicated, a lack of therapeutic benefit may be seen due to rapid drug clearance. Poor metabolizers with impaired enzyme function comprise approximately 7% of Caucasians, 2% of African Americans and 1% of Asian populations (1–3). Individualized dosage based on the knowledge of genetic constitution can be used to improve drug therapy and decrease the incidence of adverse drug effects.

A panel of assays has been developed that uses rapid PCR and hybridization probes on the LightCycler. Samples are initially genotyped for the CYP2D6*3 and CYP2D6*4 SNPs. A second set of reactions, utilizing an internally amplified reference gene and competitors, identifies the CYP2D6*5 deletion.

## Materials

LightCycler (Roche Molecular Systems, Indianapolis, USA)           *Equipment*
LightCycler capillary tubes (Roche Molecular Systems, Indianapolis, USA)

AmpliTaq (Roche Molecular Systems, Indianapolis, USA)           *Reagents*
Taqstart Antibody (Clontech, Palo Alto, USA)
LightCycler-DNA Master Hybridization Probes (Roche Molecular Systems, Indianapolis, USA)
Hybridization Probes (Idaho Technology, Salt Lake City, USA)
Primers (Peptide Core Facility-University of Utah, Salt Lake City, USA)
PicoGreen dsDNA Quantification Kit (Molecular Probes, Eugene, USA)
QIAquick PCR Purification Kit (Qiagen, Valencia, USA)

* Elaine Lyon, Ph.D., ARUP Institute for Clinical and Experimental Pathology, 500 Chipeta Way, SLC,UT, 84108, Department of Pathology, University of Utah, SLC, UT
E-mail: lyone@aruplab.com

## Procedure

DNA was prepared from peripheral blood using phenol/chloroform extraction and ethanol precipitation (4). DNA concentration was measured by spectrophotometry and adjusted to 50 ng/μL.

The primer and probe sequences are shown in Table 1. The CYP2D6–8 cluster on chromosome 22 consists of the CYP2D6 gene and two pseudogenes, CYP2D7P and CYP2D8P, which differ only slightly in base sequence from CYP2D6 (5).

**Table 1.** Primers and probes for CYP2D6*3, CYP2D6*4 and β-globin

| | Length (bp) | GC (%) | $T_m$ (°C) | Purity |
|---|---|---|---|---|
| **CYP2D6*3** | | | | |
| **Primers** | | | | |
| -CCTGAGACTTGTCCAGG | 17 | 59 | 55.7 | $A_{260}/A_{280}$=1.71 |
| -GGCCGAGAGCATACTCG | 17 | 65 | 59.4 | $A_{260}/A_{280}$=1.79 |
| **Probes** | | | | |
| -LCRed705-GCTAACTGAGCACAGG ATGAC-P | 21 | 52 | 60.7 | $C_F/C_O$=1.22 |
| -CGCTTCCAAAAGGCTTTCCTGAC CCAGCTGGATGAGC-F | 37 | 57 | 76.0 | $C_F/C_O$=1.08 |
| **CYP2D6*4** | | | | |
| **Primers** | | | | |
| -CGCCTTCGCCAACCACT (for) | 17 | 65 | 62.9 | $A_{260}/A_{280}$=1.35 |
| -GAGACTCCTCGGTCTCTC (rev) | 18 | 61 | 57.5 | $A_{260}/A_{280}$=1.61 |
| CGCCTTCGCCAACCACTCCGGTGGG TGATGGGCAGAAGGGCACAAAGCG GGAACTGGGAAGGCGGGGGACGGG GAAGGCGACCCCTTACCCGCATCTC CCACCCCCAGGATGCC (competitor) | 114 | 68 | 92.4 | $A_{260}/A_{280}$=1.76 |
| **Probes** | | | | |
| -LCRed640-CCCAACGGTCTCTTGGA CAAAGCCGTGAGCAACGTG-P | 36 | 58 | 76.3 | $C_F/C_O$=1.29 |
| -CCCAAGACGCCCCTTTCG-F | 18 | 67 | 63.2 | $C_F/C_O$=1.00 |
| **β-globin** | | | | |
| **Primers** | | | | |
| -ACACAACTGTGTTCACTAGC (for) | 20 | 45 | 58.0 | $A_{260}/A_{280}$=1.71 |
| -CAACTTCATCCACGTTCACC (rev) | 20 | 50 | 59.1 | $A_{260}/A_{280}$=1.48 |
| -CAACTTCATCCACGTTCACCTTG CCCCACAGGGTAGTAACGG (competitor) | 42 | 55 | 76.4 | $A_{260}/A_{280}$=1.42 |
| **Probes** | | | | |
| LCRed705-AGACTTCTCCTC AGGA GTCAGGTGCACCATG-P | 31 | 55 | 71.3 | $C_F/C_O$=1.39 |
| -CCACAGGGCAGTAACGG-F | 17 | 65 | 59.8 | $C_F/C_O$=0.98 |
| -CACAGGGCAGTAACG-F | 15 | 60 | 53.5 | $C_F/C_O$=1.09 |

Primers were designed to prevent concurrent amplification of the pseudogenes. CYP2D6*3 is in exon 5 and contains a single A-T base pair deletion. CYP2D6*4, spanning the intron 3/ exon 4 boundary, contains a G to A substitution. A portion of the β-globin gene was amplified as a reference gene (6).

The competitors for CYP2D6*4 and β-globin were constructed using site-directed mutagenesis PCR to introduce a single base mismatch into the wild type sequence amplicon. PCR was performed using the following master mix.

**Competitor Preparation**

Table 2. Master mix for competitor preparation

| Master Mix | Volume (μl) | [Final] |
|---|---|---|
| MgCl₂ Buffer (30 mM) | 2 | 3 mM |
| Primers (5 μM each) (reverse+competitor for CYP*4) or (forward+competitor for β-globin) | 2 | 0.5 μM each |
| dNTP (2 mM each) | 2 | 0.2 μM each |
| AmpliTaq polymerase (5 U/μl) | 0.19 | 0.05 U |
| Clontech Antibody (1.1 μg/μl) | 0.19 | 0.01 μg |
| Taq diluent | 1.62 | |
| ddH₂O | 10 | |
| Total master mix volume per reaction | 18 | |

PCR for competitor templates was performed on the LightCycler using the following conditions.

Table 3. Amplification protocol for competitor preparation

| Parameter | Value | | |
|---|---|---|---|
| Cycles | 55 | | |
| Type | None | | |
| | Segment 1 | Segment 2 | Segment 3 |
| Target temperature [°C] | 94 | 60 | 72 |
| Incubation time [s] | 2 | 5 | 20 |
| Temperature Transition rate [°C/s] | 20 | 20 | 20 |
| Acquisition mode | none | none | none |
| Adjustable gains - LC Run Version 5.32 | | | |

PCR products were purified using the QIAquick PCR Purification Kit. Competitor concentration was determined using PicoGreen and serially diluted from 500K to 1K copies/μL (7).

**LightCycler PCR**    Four LightCycler assays were developed to genotype CYP2D6*3, CYP2D6*4 and CYP2D6*5(del) of the CYP2D6 gene. The assays were performed in three tiers.

The first tier used asymmetric PCR to amplify specific regions of the CYP2D6 gene containing the *3 and the *4 polymorphisms. An aliquot of each master mix (9.0 μL) was placed in a separate capillary tube. Genomic DNA (1.0 μL of a 50 ng/μL preparation) was added to each of the two master mixes.

**Table 4.** Master mix for CYP2D6*3 and CYP2D6*4 LightCycler reactions

| Master Mix | Volume (μl) | [Final] |
|---|---|---|
| LightCycler-DNA Master Hybridization Probes (10X) | 1 | 1X |
| MgCl$_2$ (25 mM) | 0.8 | 3 mM |
| *3 or *4 primers (2:5 μM each)(forward:reverse) | 1 | 0.5 μM |
| *3 or *4 probes (2 μM each) | 1+1 | 0.2 μM |
| ddH$_2$O | 4.2 | |
| Total master mix volume per reaction | 9 | |

The second tier of reactions screened samples for the CYP2D6*5(del). CYP2D6*5(del) screening was performed using a multiplex assay comprised of the CYP2D6*3 and β-globin (acting as an internal reference gene) assays. 1.0 μL of a 50 ng/μL DNA preparation was added to a capillary containing 9.0 μL of master mix.

**Table 5.** PCR master mix for CYP2D6*5(del) screening assay

| Master Mix | Volume (μl) | [Final] |
|---|---|---|
| LightCycler-DNA Master Hybridization Probes (10X) | 1 | 1X |
| MgCl$_2$ (25 mM) | 0.8 | 3 mM |
| CYP2D6*3 primers (2:5 μM) (forward:reverse) | 1 | 0.2:0.5 μM |
| β-globin primers (5:2 μM) (forward:reverse) | 1 | 0.2:0.5 μM |
| Probes (15-mer FITC β-globin probe)(2 μM each) | 1+1+1+1 | 0.2 μM each |
| ddH$_2$O | 1.2 | |
| Total master mix volume per reaction | 9 | |

Finally, confirmation of the CYP2D6*5 (del) was performed using CYP2D6*4 and β-globin assays, incorporating internal competitors for both the target (CYP2D6*4) and reference gene (β-globin). Each master mix (8.0 μL) was placed into separate capillary tubes (non-multiplexed). DNA (1.0 μL) and 1.0 μL of the competitor (concentrations ranging from 62K copies/μL to 7.5K copies/μL) were added to the master mix.

**Table 6.** PCR master mix for CYP2D6*5(del) confirmation assay

| Master Mix | Volume (µl) | [Final] |
|---|---|---|
| LightCycler-DNA Master Hybridization Probes (10X) | 1 | 1X |
| MgCl$_2$ (25 mM) | 0.8 | 3 mM |
| CYP2D6*4 primers (2:5 µM) (forward:reverse) | 1 | 0.2:0.5 µM |
| or | or | or |
| β-globin primers (5:2 µM) (forward:reverse) | 1 | 0.2:0.5 µM |
| *4 or β-globin probes (2 µM each) | 1+1 | 0.2 µM each |
| (17-mer FITC β-globin probe) | | |
| ddH$_2$O | 3.2 | |
| Total master mix volume per reaction | 8 | |

PCR for all three tiers used the same cycling conditions.

**Table 7.** LightCycler amplification protocol for all reactions

| Parameter | Value | | |
|---|---|---|---|
| Cycles | 55 | | |
| Type | Quantification | | |
| | Segment 1 | Segment 2 | Segment 3 |
| Target temperature [°C] | 95 | 59 | 72 |
| Incubation time [s] | 3 | 10 | 15 |
| Temperature Transition rate [°C/s] | 20 | 20 | 1 |
| Acquisition mode | none | single | none |
| Automatic gain adjustment - LC Run Version 5.32 | | | |

After 55 cycles of amplification, fluorescence was monitored while the samples were heated slowly.

**Table 8.** Melting protocol for reactions

| Parameter | Value | | | |
|---|---|---|---|---|
| Cycles | 1 | | | |
| Type | Melting curve | | | |
| | Segment 1 | Segment 2 | Segment 3 | Segment 4 |
| Target temperature [°C] | 95 | 65 | 40 | 85 |
| Incubation time [s] | 2 | 60 | 120 | 0 |
| Temperature Transition rate [°C/s] | 20 | 20 | 1 | 0.1 |
| Acquisition mode | none | none | single | continuous |
| Automatic gain adjustment - LC Run Version 5.32 | | | | |

**Analysis**   Genotyping of CYP2D6*3 and CYP2D6*4 was accomplished using simple melting curve analysis (8). Polynomial estimation of the derivative curve was used with background subtraction and the sliding window set at 8 °C.

The CYP2D6*5(del) was assessed on two levels. Screening was performed by derivative curve analysis of the CYP2D6*3 and β-globin multiplexed assay. Both loci were detected in the LCred705 channel. Allele areas were calculated by the LightCycler. The ratio of the β-globin:CYP2D6*3 allele area was calculated and normalized using the allele area ratio generated by a known wild type DNA. An allele area ratio between 1.24 and 0.76 generated from a wild type CYP2D6*3 suggests a single copy of the CYP2D6 gene. Ratios greater than 1.24 suggest a deletion of the CYP2D6 gene.

Confirmation of suspected deletion samples was accomplished using two competitive PCR assays. The two assays, CYP2D6*4 and β-globin, incorporated the use of internal competitors, differing from their respective wild type sequences by a single base mismatch under the reporter probe. CYP2D6*4 was analyzed in the LCred640 channel and β-globin was analyzed in the LCred705 channel. Competitors were serially diluted and amplified with genomic DNA to identify competitor concentrations with near equal peak areas to known wild type CYP2D6 genomic DNA (9). The identified competitor range was then amplified with the unknown samples. A ratio of competitor to target (C/T) peak area was calculated for each competitor concentration for both the CYP2D6*4 and β-globin loci. The final result was obtained by dividing the CYP2D6*4 C/T ratio by the β-globin C/T ratio.

## Results

$T_m$'s for all alleles are listed in Table 9.

**Table 9.** Melting Temperature Table

| Gene | Allele | Pairing | $T_m$ (observed) °C |
|---|---|---|---|
| CYP2D6*3 | wild type | A-T match | 65.5 |
|  | *3 polymorphism | A deletion | 60.7 |
| CYP2D6*4 | wild type | A-C mismatch | 59.6 |
|  | *4 polymorphism | A-T match | 66.5 |
|  | competitor | A-C and C-A mismatches | 49.1 |
| β-globin | wild type (17-mer probe) | G-C match | 61.9 |
|  | wild type (15-mer probe) | G-C match | 54.1 |
|  | competitor | A-C mismatch | 52.3 |

Genotyping results of the CYP2D6*3 and the CYP2D6*4 were used in the interpretation of the two CYP2D6*5(del) assays.

Figure 1 illustrates three different genotypes using the CYP2D6*5(del) screening method. (1) A homozygous wild type at the CYP2D6*3 locus with a normalized β-globin:wild type ratio of 1.00. (2) A heterozygous CYP2D6*3 / wild type with a normalized β-globin:wild type ratio of 1.90. (3) A heterozygous CYP2D6*5(del) / wild type with a normalized β-globin:wild type ratio of 1.67.

Suspected deletion samples were confirmed using the β-globin and CYP2D6*4 competitive assays. Three serially diluted competitor concentrations were amplified with each genomic DNA for both the CYP2D6*4 and the β-globin locus. Figure 2 illustrates a homozygous wild type and a CYP2D6*5(del) / wild type sample. The homozygous wild type sample maintained a normalized CYP2D6*4:

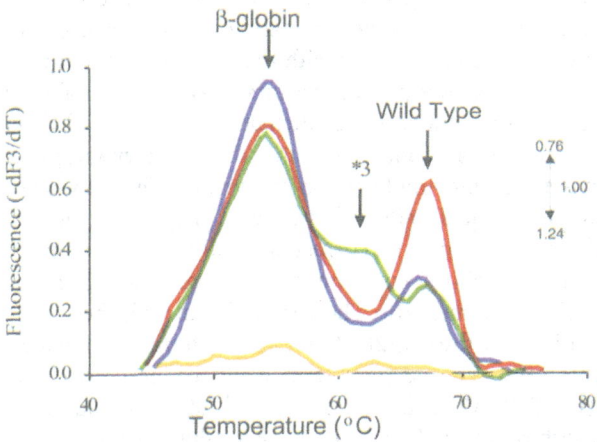

**Fig. 1.** Homozygous wild type (—) produces a normalized β-globin:wild type peak area ratio of 1.00. Heterozygous *5(del)/wild type (—) ratio equals 1.67. Heterozygous *3/wild type (—) shows three distinct peaks, with a normalized β-globin:wild type peak area ratio of 1.90. No template control (—)

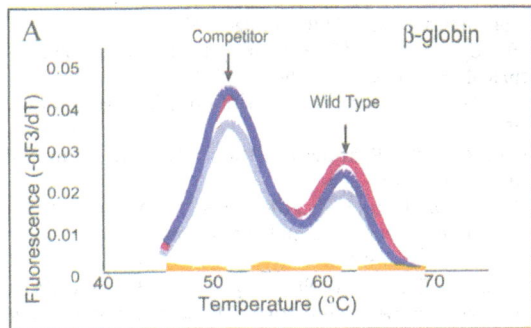
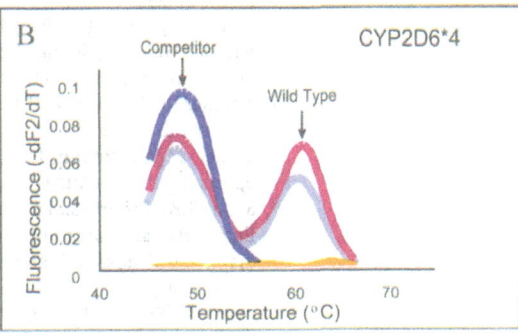

**Fig. 2. A.** Homozygous wild type DNA (—) and (—) co-amplified with β-globin competitor (initial concentration = 31K copies). CYP2D6*5(del) DNA (—) co-amplified with β-globin competitor (initial concentration = 1K copies). No template control (—). **B.** Homozygous wild type DNA (—) and (—) co-amplified with CYP2D6*4 competitor (initial concentration = 62K copies). CYP2D6*5(del) DNA (—) co-amplified with CYP2D6*4 competitor (initial concentration = 1.9K copies). No template control (—)

β-globin ratio of 1.0. The CYP2D6*5(del) / wild type sample was identifed by the lack of a genomic peak when tested with the CYP2D6*4 assay, even with decreasing the competitor concentration to 2K copies/µL, while the β-globin assay maintained a relative 1:1 C/T peak area ratio.

## Comments

As illustrated by the CYP2D6*5(del) system, gene deletions are detected utilizing two different methods: co-amplification of a target and internal reference gene and amplification of competitors in conjunction with the target and reference gene. Amplification of an internal reference gene multiplexed with the target gene provides a reliable and rapid means of screening. Internal competitors add an additional level of complexity and control and are valuable for confirmation of suspected deletions. The addition of the competitors acts both to quantify the unknown sample as well as to control for sample to sample amplification efficiency differences, since both the competitor and sample are equally affected by the efficiency of the reaction.

Competitor concentrations may need to be corrected for differing amounts of input genomic DNA. Using the β-globin locus, competitor concentrations can be adjusted to give a 1:1 competitor:genomic ratio. The same adjustment is then made in the competitor concentrations for the CYP2D6*4 locus to determine a normalized wild type:polymorphic ratio. This same system could be used to detect CYP2D6 duplicated allele (CYP2D6*X2). Allele ratios for duplications would be the converse of deleted genes, with ratios <0.75 indicating a gene duplication.

## References

1. Kalow, W., Grant, D.M. Pharmacogenetics. In: Scriver C.R., Beaudet A.L., Sly W.S., Valle D., eds. The Metabolic and Molecular *Bases* of Inherited Disease. New York: McGraw-Hill, Inc., 2001:225–255
2. Wolf, C.R., Smith, G. (1999) Pharmacogenetics. Br Med Bull; 55(2):366–386
3. Van der Weide, J., Steijns, L.S.W. (1999) Cytochrome P450 enzyme system: genetic polymorphisms and impact on clinical pharmacology. Ann Clin Biochem, 36:722–729
4. Thomas, S.M., Moreno, R.F., Tilzer, L.L. (1989) DNA extraction with organic solvents in gel barrier tubes. Nucleic Acids Res., 17:5411
5. Linder, M.W., Prough, R.A., Valdes, Jr. R. (1997) Pharmacogenetics: a laboratory tool for optimizing therapeutic efficiency. Clin Chem., 43:254–266
6. Saiki, R.K., Scharf, S., Faloona, F., Mullis, K.B., Horn, G.T., Erlich, H.A., Arnheim, N. (1985) Enzymatic amplification of β-globin genomic sequences and restriction site analysis for diagnosis of sickle cell anemia. Science, 230:1350–1354
7. Millson, A., Suli, A., Hartung, L., Kunitake, S., Bennett, A., Nordberg, M.C.L., Hanna, W., Wittwer, C.T., Seth, A., Lyon, E. (2003) Comparison of two quantitative PCR methods for detecting HER2/*neu* amplification. J Mol Diagn. 5(3):184–190
8. Lay, M.J., and Wittwer, C.T. (1997) Real-time fluorescence genotyping of factor V Leiden during rapid-cycle PCR. Clin. Chem., 43:2262–2267
9. Lyon, E., Millson, A., Lowery, M.C., Woods, R., Wittwer, C.T. (2001) Quantification of HER2/*neu* gene amplification by competitive PCR using fluorescent melting curve analysis. Clin. Chem., 47:844–851

# Trisomy 21 Detected by SNP Allele Ratios

Genevieve Pont-Kingdon*, Elaine Lyon

## Introduction

The controlled balance of gene expression in eukraryotic cells depends on gene dosage. Additional genetic material may result from single gene duplication, from gene amplification, or from whole chromosome replication. Gene duplication and amplification have been linked to oncogenic transformation while chromosomal aneuploidies are a common cause of severe birth defects and lost pregnancies.

Trisomy 21 is the most common chromosomal aneuploidy compatible with life and the most common cause of mental retardation, occurring in 1 per 800 births. Like other aneuploidies, trisomy 21 is typically analyzed by cytogenetics or fluorescence *in situ* hybridization.

Recently, several molecular techniques have been developed for rapid detection of trisomies.

Relative quantitative PCR using chromosome 21 genes and reference genes on other chromosomes can be used to assess gene (chromosome) copy number [1]. Alternatively, quantification of alleles at polymorphic loci on chromosome 21 can distinguish the presence of 2 (normal) from 3 (trisomy) alleles. Highly-variable small tandem repeats (STR) in PCR products can be detected by length polymorphisms and quantified [2–5].

Alternatively, single nucleotide polymorphisms (SNPs) can be detected and quantified by melting analysis for trisomy 21 diagnosis [6]. Melting curves have also been used to quantify alleles in gene duplication [7], and in gene amplification in the presence of a competitor [8]. A panel of six SNPs along chromosome 21 that are heterozygous in 30 to 50% of the population was selected. Normalized melting curve allele ratios of trisomy 21 samples are approximately 0.5 or 2.0 depending on which allele is over-represented.

## Materials

LightCycler® instrument (Roche Diagnostics, Mannheim, Germany)     Instrument
MagNA Pure LC instrument (Roche Applied Science, Mannheim Germany)

---

* Genevieve Pont-Kingdon, ARUP Institute for Clinical and Experimental Pathology, 500 Chipeta Way, Salt Lake City, Utah 84108
E-mail: pontkig@aruplab.com

Reagents Primers (DNA-Peptide Core facility: University of Utah)
Probes: Fluorescein, LC Red640, and LC Red705 (Idaho Technology: Salt Lake City, Utah USA)
LightCycler DNA Master Hybridization Probes (Roche Diagnostics, Mannheim, Germany)
PUREGENE® DNA Isolation Kit (Gentra)
1X PBS (Ambion)
MagNA Pure LC DNA Isolation Kit I (Roche Applied Science, Mannheim Germany)

## Procedure

**Samples**

DNA from 3 to 5 million cultured amniocytes (fixed in methanol-acetic acid; 3:1) were extracted with a PUREGENE DNA Isolation Kit after two successive washes were performed with 1X PBS. DNA from fresh or whole blood was extracted on the MagNA Pure LC instrument using the MagNA Pure LC DNA Isolation Kit I. DNA was eluted in 20–50 µl.

**Primers and Probe Selection**

Primer and probe sequences are presented in Table 1, and have been previously reported [6]. Probes were designed to ensure the highest possible $T_m$ difference (delta $T_m$) between both alleles of each SNP.

**LightCycler PCR**

For each sample, three reactions are performed (Table 2). The first uses master mix 1 (MM1), and co-amplifies and analyzes WIAF 2643 and WIAF 899. The second (MM2) co-amplifies and analyzes WIAF 1538 and WIAF 2215. The third (MM3) co-amplifies and analyzes WIAF 1882 and WIAF 1943. For each reaction, 1 µl of DNA (approximately 50 ng) was added to 9 µl of MM in a final reaction volume of 10 ul.

Additionally, at each run, a known heterozygous diploid was amplified with the same MM. Melting curve areas calculated from the diploid sample were used to normalize the data.

**Table 1.** PCR primers and probes for 6 SNPs on chromosome 21

| WIAF-2643 | | | | |
|---|---|---|---|---|
| Sequence | Length | GC (%) | $T_m$ (°C) | Purity |
| Primers | | | | |
| AACCCAGTGTGGGAGGAGAA | 20 | 55.0 | 63.2 | 1.81 |
| GTGGTGCTGTGGGGCTAG | 18 | 66.7 | 63.3 | 1.72 |
| Probes | | | | |
| CAGAATAAATAGAACAGTAGAATG-F | 24 | 29.2 | 51.5 | 1.25 |
| LCRed705-TCACAGATGGGTAATTACACATG TAAATGAGCTC-P | 34 | 38.2 | 64.3 | 0.92 |

**Table 1.** *Continued*

| WIAF-899 | | | | |
| --- | --- | --- | --- | --- |
| Sequence | Length | GC (%) | T$_m$ (°C) | Purity |
| **Primers** | | | | |
| CAGGCAGGACTTCAGTGTCA | 20 | 55.0 | 62.3 | 1.74 |
| GTCATCTGGGACAGGTCACC | 20 | 60.0 | 62.4 | 1.73 |
| **Probes** | | | | |
| TTCCTGTTCCACGAAGAGGAC-F | 21 | 52.4 | 60.1 | 0.92 |
| LCRed640-TTTTGTTCACAATTGGATCACAAT GCAGAGGAGTCTGTT-P | 39 | 38.5 | 68.5 | 0.72 |

| WIAF-1538 | | | | |
| --- | --- | --- | --- | --- |
| **Primers** | | | | |
| TGTTTGTGTTCCAGCCACAT | 20 | 45 | 60.9 | 1.39 |
| CTCTCAGTTAGCAGCTGGGC | 20 | 60 | 63.0 | 1.41 |
| **Probes** | | | | |
| LCRed705-CCAATGTTATGTCGAAACTGCATT GTAAAAAG-P | 32 | 34 | 62.4 | 0.91 |
| GCGCACCATTCATCATTTAGGCTTGTGGTTTG TTGTTTACTCT-F | 43 | 42 | 71.2 | 0.51 |

| WIAF-2215 | | | | |
| --- | --- | --- | --- | --- |
| **Primers** | | | | |
| GGCTCACAAACATCCAC | 17 | 53 | 56.2 | 1.84 |
| CATCAAAGCACCTGTCG | 17 | 53 | 56.5 | 1.58 |
| **Probes** | | | | |
| TGGTCCCCCTGCCGAGGG-F | 18 | 78 | 66.8 | 0.59 |
| LCRed640-GTGCGGCCTCTGCAAGGTTCGGG GGTGGCTTCGTTTGCCTGG-P | 42 | 67 | 81.1 | 1.23 |

| WIAF-1882 | | | | |
| --- | --- | --- | --- | --- |
| **Primers** | | | | |
| TTTTTTGGCTTGTCTGCAGA | 20 | 40 | 58.9 | 1.59 |
| CAGTGAGCCAGCACTCTTGG | 20 | 60 | 63.6 | 1.62 |
| **Probes** | | | | |
| LCRed640-GAGGCAGCGCTTACAGGAG-P | 19 | 63 | 61.2 | 1.50 |
| CCCAAGTGCACACTAGGCAATGTAAGCTCC-F | 30 | 47 | 68.7 | 0.80 |

| WIAF-1943 | | | | |
| --- | --- | --- | --- | --- |
| **Primers** | | | | |
| TTTTTAACGAAATCTCACTACTGCA | 25 | 32 | 59.2 | 1.83 |
| CTATGCACCATGTACTGTTCTAAGC | 25 | 44 | 61.4 | 1.59 |
| **Probes** | | | | |
| GCTAATGAATGCACAGAGTAT-F | 21 | 38 | 54.3 | 1.12 |
| LCRed705-GCCTGCAAAATAATAATTGAGATT CTATTTTTAAG-P | 35 | 26 | 59.2 | 0.61 |

**Table 2.** Master mixes used for allele quantification at the 6 SNP loci

| Master Mixes | Volume [µl] | [Final] |
|---|---|---|
| **MM1** | | |
| Primer/probe premix-WIAF-2643 | 1 | 2643-primers: 0.5 µM each |
| (5 µM each, 2 µM each) | | 2643-probes: 0.2 µM each |
| Primer/probe premix-WIAF-899 | 1 | 899-primers: 0.5 µM each |
| (5 µM each, 2 µM each) | | 899-probes: 0.2 µM each |
| LightCycler DNA Master Hybridization Probes | 1 | 1x |
| MgCl$_2$ (25 mM) | 0.8 | 3 mM |
| Water | 5.2 | |
| Total Master Mix volume per reaction | | 9 |
| **MM2** | | |
| Primer/probe premix-WIAF-1538 | 1 | 1538-primers: 0.5 µM each |
| (5 µM each, 2 µM each) | | 1538-probes: 0.2 µM each |
| Primer/probe premix-WIAF-2215 | 1 | 2215-primers: 0.5 µM each |
| (5 µM each, 2 µM each) | | 2215-probes: 0.2 µM each |
| LightCycler DNA Master Hybridization Probes | 1 | 1x |
| MgCl$_2$ (25 mM) | 0.8 | 3 mM |
| Water | 5.2 | |
| Total Master Mix volume per reaction | | 9 |
| **MM3** | | |
| Primer/probe premix-WIAF-1882 | 1 | 1882-primers: 0.5 µM each |
| (5 µM each, 2 µM each) | | 1882-probes: 0.2 µM each |
| Primer/probe premix-WIAF-1943 | 1 | 1943-primers: 0.5 µM each |
| (5 µM each, 2 µM each) | | 1943-probes: 0.2 µM each |
| LightCycler DNA Master Hybridization Probes | 1 | 1x |
| MgCl$_2$ (25 mM) | 0.8 | 3 mM |
| Water | 5.2 | |
| Total Master Mix volume per reaction | | 9 |

The following parameters were used for amplification and melting.

**Table 3.** Amplification and melting parameters

| Amplification Segment Number | Temperature Target (°C) | Cycles Hold Time (sec) | 35 Slope (C°/sec) | Acquisition Mode |
|---|---|---|---|---|
| 1 | 95 | 2 | 20 | None |
| 2 | 60 | 10 | 20 | Single |
| 3 | 72 | 15 | 2 | None |
| First Melting Curve | | Cycles | 1 | |
| 1 | 95 | 0 | 20 | None |
| 2 | 35 | 120 | 20 | None |
| 3 | 85 | 0 | 0.1 | Continuous |
| Second Melting Curve | | Cycles | 1 | |
| 1 | 95 | 0 | 20 | None |
| 2 | 35 | 120 | 20 | None |
| 3 | 85 | 0 | 0.2 | Continuous |
| Third Melting Curve | | Cycles | 1 | |
| 1 | 95 | 0 | 20 | None |
| 2 | 35 | 120 | 20 | None |
| 3 | 85 | 0 | 0.3 | Continuous |

LightCycler software, version 5.32 was used with automatic gain adjustment

Analysis was performed using color compensation to correct for fluorescence overlap between channels.

Melting curves were analyzed using polynomial averaging over 8°C intervals and LightCycler software, version 3.5 with the following conditions:

**Table 4.** Analysis parameters for each SNP

| SNP | Channel | Melting Rate (Co/sec) | Temperature Window (°C) |
| --- | --- | --- | --- |
| WIAF-2643 | F2 | 0.2 | 40–66 |
| WIAF-899 | F3 | 0.1 or 0.2 | 46–73 |
| WIAF-1538 | F2 | 0.2 | 54–72 |
| WIAF-2215 | F3 | 0.1 | 47–77 |
| WIAF-1882 | F2 | 0.2 | 50–75 |
| WIAF-1943 | F3 | 0.1 | 45–65 |

The areas of least and most stable alleles were determined by non-linear least sequences fitting of multiple gaussians (LightCycler software). The curve area ratios were normalized to the ratio of heterozygous diploid samples.

## Result and Discussion

For each SNP, characteristic $T_m$ and delta $T_m$ values identify both alleles. As expected, the melting curves $T_m$s are not different between trisomic (3N) and diploid (2N) samples (Table 5). The $T_m$s are also identical regardless of the sample type, i.e., DNA extracted from fixed cells or from whole blood (Table 5).

**Table 5.** Mean melting temperature of each allele in diploid (2N) and trisomic (3N) samples

| Loci | Alleles | Sample Types | Perfect Match | $T_m$ Mismatch | Delta $T_m$ |
| --- | --- | --- | --- | --- | --- |
| WIAF 2643 | G/C | 3N Fixed cells | 57.5 | 48.6 | 8.9 |
|  |  | 2N whole blood | 58.1 | 49.4 | 8.7 |
| WIAF 899 | C/T | 3N Fixed cells | 63.4 | 55.0 | 8.7 |
|  |  | 2N whole blood | 63.5 | 55.1 | 8.4 |
| WIAF 1538 | A/G | 3N Fixed cells | 67.7 | 61.7 | 5.9 |
|  |  | 2N whole blood | 67.6 | 61.7 | 5.9 |
| WIAF 2215 | G/A | 3N Fixed cells | 69.1 | 59.1 | 10.0 |
|  |  | 2N whole blood | 68.9 | 59.8 | 9.1 |
| WIAF 1882 | A/C | 3N Fixed cells | 67.7 | 57.4 | 10.3 |
|  |  | 2N whole blood | 66.4 | 57.1 | 10.7 |
| WIAF 1943 | T/C | 3N Fixed cells | 60.1 | 51.3 | 8.9 |
|  |  | 2N whole blood | 60.0 | 51.2 | 8.9 |

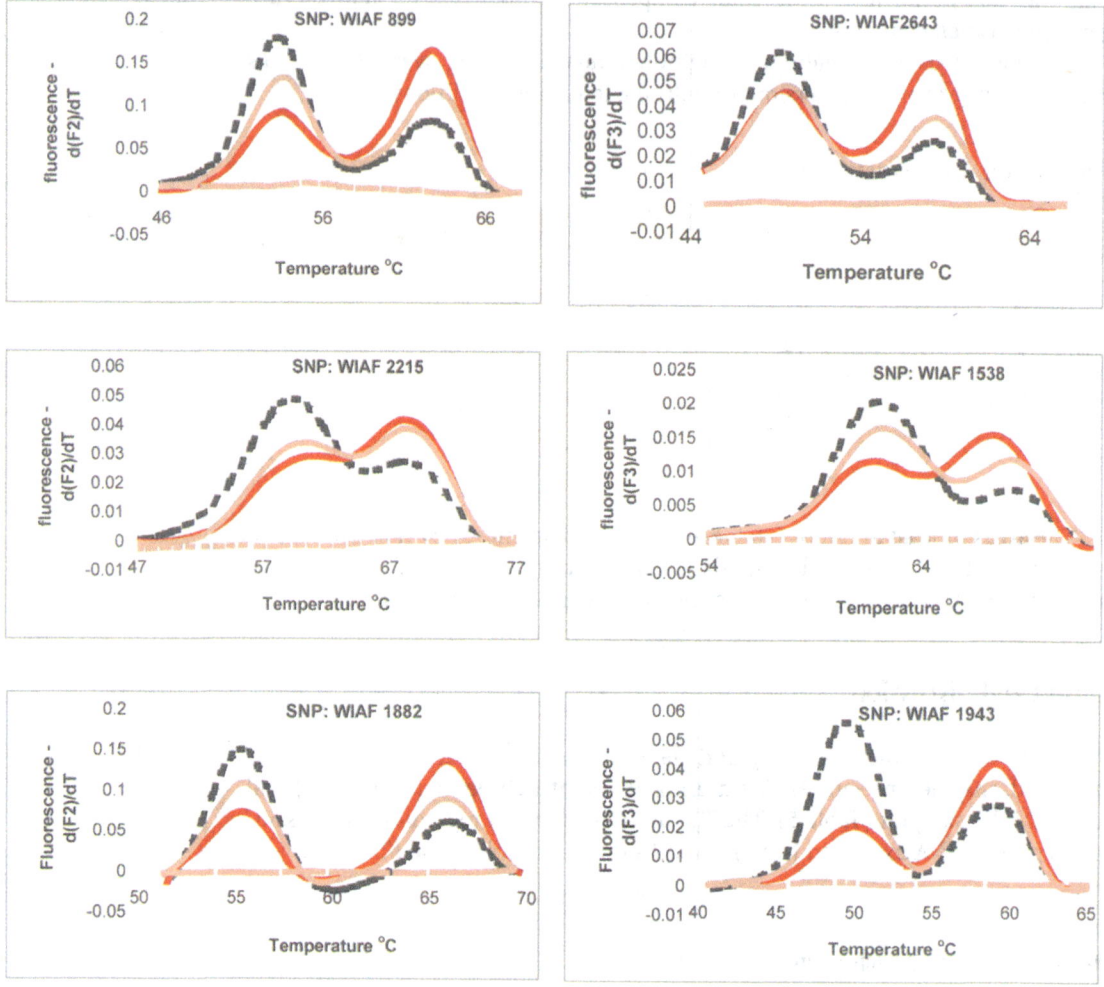

**Fig. 1.** Melting curves for the six SNPs from heterozygous diploid and trisomic samples. (Reproduced from [6] with permission of AACC). Derivative melting curves of trisomic samples are shown by the dashed black (overrepresentation of the less stable allele) and red lines (overrepresentation of the most stable allele); salmon line, nontrisomic reference; dashed salmon line, no-template control

Melting curve examples for each of the six SNP loci from diploid and trisomic samples are shown in Figure 1.

**Melting Curve Ratios**    The mean (±SD) of the normalized allele ratios for each SNP are listed in Table 6. These data were obtained from duplicate measurements and are presented in detail elsewhere [6]. Diploid samples had a normalized allele ratio around 1.00

**Table 6.** Allele ratios for diploid and trisomic samples (mean ± standard deviation)

| SNP | Diploid | Trisomic (least stable allele) | Trisomic (most stable allele) |
|---|---|---|---|
| WIAF 2643 | 1.08 (0.12) | 0.61 (0.03) | 1.90 (0.14) |
| WIAF 899 | 1.01 (0.11) | 0.51 (0.08) | 2.03 (0.41) |
| WIAF 1538 | 1.00 (0.11) | 0.55 (0.08) | 1.86 (0.10) |
| WIAF 2215 | 1.98 (0.13) | 0.60 (0.09) | 1.86 (0.25) |
| WIAF 1882 | 1.09 (0.14) | 0.51 (0.06) | 2.09 (0.34) |
| WIAF 1943 | 0.96 (0.15) | 0.46 (0.13) | 1.84 (0.33) |

(range 0.96 to 1.09) while trisomic samples had ratios around 0.5 (range 0.46 to 0.61) or 2.00 (range 1.84 to 2.09).

## Comments

The main factors used to select SNPs were the heterozygosity index (percent of heterozygous individuals in a random population) and the melting curve quality (a diploid allele ratio close to 1 and delta $T_m$ >5°C). To detect partial trisomy 21 of the down syndrome critical region [9, 10], two of the SNPs were from this region.

*Selection of SNPs*

Melting conditions had a profound effect on the allele areas. In general, slower melting at a rate of 0.1°C/s increased the area of the least stable allele, and therefore, increased the allele-1/allele-2 ratios. For each SNP, data were acquired at three different melting rates, and the rate that resulted in an allele area ratio closest to 1.0 for diploid samples was used.

*Melting Conditions and Allele Ratios*

This method can be applied to any gene dosage defect where genes are present in a 2:1 ratio. These conditions include gene duplication [7] and aneuploidies [6]. Of these, trisomies of chromosomes 13, 18 and X are compatible with life and are tested by cytogenetics or molecular techniques [5]. A panel of SNPs in these chromosomes can be easily designed. Gene dosage analysis of several chromosomes would require analysis of many SNPs. Such high throughput needs can be partly met by multiplexing fluorophores and/or $T_m$ [11]. Alternatively, instruments that allow post-PCR product analysis by melting in 96- or 384-well format, such as the LightTyper (Roche), would be ideal for this type of analysis.

*Other Applications*

## References

1. Zimmermann B, Holzgreve W, Wenzel F, Hahn S. Novel real-time quantitative PCR test for trisomy 21. Clin Chem 2002;48:362–3
2. Mansfield ES. Diagnosis of Down syndrome and other aneuploidies using quantitative polymerase chain reaction and small tandem repeat polymorphisms. Hum Mol Genet 1993;2:43–50

3. Verma L, Macdonald F, Leedham P, McConachie M, Dhanjal S, Hulten M. Rapid and simple prenatal DNA diagnosis of Down's syndrome. Lancet 1998;352:9–12

4. Pertl B, Kopp S, Kroisel PM, Hausler M, Sherlock J, Winter R, et al. Quantitative fluorescence polymerase chain reaction for the rapid prenatal detection of common aneuploidies and fetal sex. Am J Obstet Gynecol 1997;177:899–906

5. Findlay I, Toth T, Matthews P, Marton T, Quirke P, Papp Z. Rapid trisomy diagnosis (21, 18, and 13) using fluorescent PCR and short tandem repeats: applications for prenatal diagnosis and preimplantation genetic diagnosis. J Assist Reprod Genet 1998;15:266–75

6. Pont-Kingdon G, Lyon E. Rapid detection of aneuploidy (trisomy 21) by allele quantification combined with melting curves analysis of single-nucleotide polymorphism loci. Clin Chem 2003;49:1087–94

7. Ruiz-Ponte C, Loidi L, Vega A, Carracedo A, Barros F. Rapid real-time fluorescent PCR gene dosage test for the diagnosis of DNA duplications and deletions. Clin Chem 2000;46:1574–82

8. Millson A, Suli A, Hartung L, Kunitake S, Bennet A, Lowry-Nordberg MC, et al. Comparison of two Quantitative PCR methods for Detecting HER2/neu Amplification. J Mol Diag 2003;5:184–90

9. Korenberg JR, Chen XN, Schipper R, Sun Z, Gonsky R, Gerwehr S, et al. Down syndrome phenotypes: the consequences of chromosomal imbalance. Proc Natl Acad Sci USA 1994; 91:4997–5001

10. Sinet PM, Theophile D, Rahmani Z, Chettouh Z, Blouin JL, Prieur M, et al. Mapping of the Down syndrome phenotype on chromosome 21 at the molecular level. Biomed Pharmacother 1994;48:247–52

11. Bernard PS, Pritham GH, Wittwer CT. Color multiplexing hybridization probes using the apolipoprotein E locus as a model system for genotyping. Anal Biochem 1999;273:221–8

# Quantitative Chimerism Analysis by Allele-Specific Real-Time PCR of a 10 bp Insertion/Deletion Polymorphism within the Promotor Region of Factor VIIc

Hendrik Reuter, Björn Tews, Jochen Wilhelm, Meinhard Hahn*

## Introduction

Chimerism analysis is a valuable tool in medical diagnostics, population genetics, and food control. The selective quantification of one allele in samples containing higher amounts of very similar sequences can be achieved by several different techniques. Immunoglobulin and T-cell receptor genes, Y-chromosomal sequences [1], restriction fragment length polymorphisms (RFLP), short tandem repeats (STR), and variable number of tandem repeats (VNTR) [2] are chosen as typical targets that have been selectively quantified in PCR-based chimerism analysis. Genetic length variations and RFLPs [3] can be identified by fragment length, but quantification must be performed after electrophoretic separation of the PCR products, *e.g.*, by capillary electrophoresis [4]. Unfortunately, the dynamic range of this time consuming method is approximately 1:20, *i.e.*, not less than one copy of one allele among 20 copies of another. The highly underpopulated allele is not amplified to detectable amounts before the plateau phase of the PCR is reached.

For sensitive chimerism analysis, we prefer a combination of rapid-cycle real-time PCR [5] and allele-specific amplification using short insertion/deletion sequences as markers. This method allows selective quantification of one allele present in the sample. The non-amplified alleles do not interfere with amplification or quantification. The dynamic range is limited by the probability of amplifying related sequences. This probability is dependent on the sequence variation between the alleles, the primer sequence, and the polymerase used. The discrimination ratio compares the lowest detectable amount of the amplified allele, to the lowest detectable amount of the non-amplified allele that gives a false-positive amplification. To demonstrate this technique, we chose a 10-bp insertion/deletion variation within the promotor region of the *factor VIIc* (*F-VIIc*) gene [6]. This polymorphism is associated with clinical risk factors [7, 8] and demonstrates that chimerism analysis by real-time PCR is simple, fast, and inexpensive.

* Meinhard Hahn, Deutsches Krebsforschungszentrum, Abteilung Molekulare Genetik, B060, Im Neuenheimer Feld 280, D-69120 Heidelberg, Germany, E-mail: m.hahn@dkfz.de

## Materials

**Equipment**

LightCycler® instrument (Roche Diagnostics, Mannheim, Germany)
LightCycler® centrifuge adaptors and capillaries (Roche Diagnostics, Mannheim Germany)
LightCycler® software, version 3.01 (Roche Diagnostics, Mannheim Germany)
OLIGO Primer Analysis software, version 5.0 (National Biosciences, Plymouth, Minnesota USA)
*So*FAR analysis software (own development) [9, 10]
U-3000 UV spectrophotometer (Hitachi/Colora, Lorch, Germany)
ABI PRISM® 310 Genetic Analyzer (Applied Biosystems, Weiterstadt, Germany)
Collection software, version 1.0.4 (Applied Biosystems, Weiterstadt, Germany)
GeneScan® software, version 2.1 (Applied Biosystems, Weiterstadt, Germany)
GS STR POP4 module (Applied Biosystems, Weiterstadt, Germany)
GeneAmp® PCR System 2400 Thermal Cycler (Applied Biosystems, Weiterstadt, Germany)

**Reagents**

QIAamp® DNA Blood reagent set (QIAGEN, Hilden, Germany)
10 X PCR buffer (100 mM Tris-HCl, 15 mM MgCl$_2$, 500 mM KCl, pH 8.3 at 20°C)
PCR primers, HPSF grade (MWG, Ebersberg, Germany)
5´- FAM-labeled primer (Applied Biosystems, Weiterstadt, Germany)
dATP, dCTP, dGTP and dTTP (Roche Diagnostics, Mannheim, Germany)
MgCl$_2$, 25 mM (Roche Diagnostics, Mannheim, Germany)
Bovine serum albumin (BSA), molecular biology grade (Roche Diagnostics, Mannheim, Germany)
*Taq* DNA polymerase, recombinant (5 U/µl) (Roche Diagnostics, Mannheim, Germany)
SYBR Green I (Roche Diagnostics, Mannheim, Germany)
Reagents for ABI PRISM® 310 Genetic Analyzer (Applied Biosystems, Weiterstadt, Germany)
POP4 polymer (Applied Biosystems, Weiterstadt, Germany)
TSR (template suppression reagent) (Applied Biosystems, Weiterstadt, Germany)

## Procedure

**DNA Isolation**

Human genomic RNA-free DNA was isolated from lymphocytes of fresh or previously frozen EDTA-treated blood using the QIAamp DNA Blood Kit following the manufacturer's instructions, including the RNase A digestion step. DNA quality and concentration were evaluated by recording UV absorption spectra between 220 and 320 nm. DNA concentrations were calculated using the relation 1 OD$^{260 nm}$ = 50 µg/ml.

**Oligonucleotides**

The primer sequences for *F-VIIc* are shown in Table 1; their relative positions within the gene are shown in Figure 1. They were checked with OLIGO Primer

**Table 1.** Primer sequences for F-VIIc

| Name | Sequence (5′–3′) | Location | Length | Purity | GC (%) | $T_m$ (°C) |
|------|------------------|----------|--------|--------|--------|------------|
| H1 | GTC TGG AGG CTC TCT TCA AA | 77–96 | 20 | 1.80 | 50 | 64.8 |
| H1-FAM | FAM-GTC TGG AGG CTC TCT TCA AA | 77–96 | 20 | 0.81 | 50 | 64.8 |
| R1 | AGA ATT CCA AAC CCC TAA TG | 77–96 | 20 | 1.79 | 38 | 60.8 |
| R2 | AAT TCC AAA CCC CTA <u>GGA</u> <u>TA</u> | 195–224 | 20 | 1.80 | 40 | 61.1 |
| R3 | GGC ACC AAC ACT TCA AAT AC | 203–222 | 20 | 1.85 | 45 | 62.9 |

The sequence complementary to the insertion is underlined

Analysis software, version 5.0 for absence of false priming sites, formation of primer dimers, and secondary structure.

**LightCycler PCR**

Initially, genomic DNA samples were denatured by boiling in a water bath for 10 min. Two separate amplifications were performed, one using primer pair H1/R1, the other using primer pair H1/R2. Eight µl of the following master mix (either with H1/R1 or H1/R2 primer mixture) was added to each 2-µl sample (containing 10–30 ng DNA).

**Table 2.** LightCyler PCR master mix table

| Master Mix | Volume [µl] | [Final] |
|------------|-------------|---------|
| PCR buffer (10×) | 1.0 µl | 1× |
| BSA (5 mg/ml) | 1.0 µl | 0.5 mg/ml |
| Primer pair (5 µM of each primer; either H1/R1 or H1/R2) | 1.0 µl | 0.5 µM each |
| dNTPs (2 mM each) | 1.0 µl | 0.2 mM each |
| MgCl$_2$ (25 mM) | 1.0 µl | 4 mM |
| SYBR-Green I (1:1000) | 0.1 µl | $10^{-5}$ dilution |
| *Taq* DNA polymerase (5 U/µl) | 0.1 µl | 0.5 U |
| H$_2$O (PCR-grade) | 2.8 µl | |
| Total master mix volume per reaction | 8 µl | |

The reaction mixture was mixed at 4°C to limit unspecific products (*e.g.*, primer dimers), and was distributed into the pre-chilled glass capillary cuvettes. Sealing was followed by a brief centrifugation step.

The following amplification protocol was used:

**Table 3.** LightCyler PCR amplification protocol

| Parameter | Value | | |
|---|---|---|---|
| Cycles | 50 | | |
| Type | Amplification | | |
| | Segment 1 | Segment 2 | Segment 3 |
| Target temperature [°C] | 95 | 65 | 72 |
| Incubation time [s] | 0 | 5 | 10 |
| Temperature transition rate [°C/s] | 20 | 20 | 20 |
| Acquisition mode | None | Single | None |
| Gains | F1=5 | | |

**Standard PCR** For comparison and verification of allelotypes the *F-VIIc* polymorphism was analyzed by capillary electrophoresis with laser induced fluorescence detection, an established method that discriminates alleles *via* different fragment lengths. Each DNA sample was amplified by conventional PCR in a volume of 25 µl containing 0.5 µM of the dye labeled forward primer H1-FAM, the unlabeled reverse primer R3, 0.2 mM of each dATP, dCTP, dGTP and dTTP, 1.5 U *Taq* DNA polymerase, 1.5 mM $MgCl_2$ and 20–30 ng of human genomic DNA in 1 x PCR-buffer. After an initial denaturation step for 180 s at 95°C, 35 cycles of amplification with 30 s at 95°C, 30 s at 55°C, and 60 s at 72°C followed.

**Capillary Electrophoresis** After amplification, each reaction was diluted 40-fold with water. An aliquot of 2 µl of each dilution was added to 12 µl of TSR, denatured for 3 min at 95°C, and immediately chilled on ice. Sample injection was done electrokinetically for 5 s at 15 kV. DNA fragments were separated at 15 kV and 60°C in a 47 cm x 50 µm capillary of uncoated fused silica filled with polymer POP4 using the ABI PRISM 310 Genetic Analyzer with module ´GS STR POP4 (1 ml) C´. The laser induced fluorescence signals were recorded by the collection software, version 1.0.4 and evaluated with GeneScan software, version 2.1 as described [11].

## Results

**Primer Design** Two *F-VIIc* allele-specific primer pairs (H1/R2 for the allele containing the 10-bp insertion [+] and H1/R1for the allele lacking the insertion [-]) and a third generic primer-set (H1/R3 for capillary electrophoresis) were designed based on the human gene *F-VIIc* sequence (Figure 1, GenBank accession no. J02933). To achieve maximal discrimination between the two alleles, primer R1 – specific for the (-) allele – crosses the deletion site with its 3´-terminal part. In contrast, primer R2 is specific for the (+) allele, with its 3´-terminus complementary to six nucleotides of the 10-bp insertion. Hence, primer pair H1/R1 amplifies the (-)-allele but not the (+)-allele, and *vice versa* for primer H1/R2.

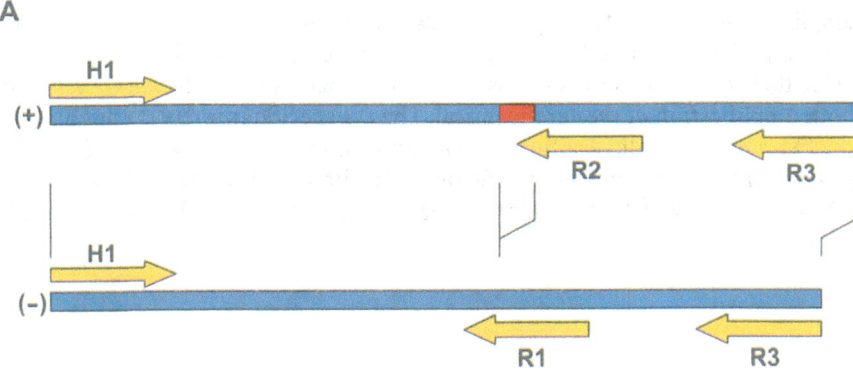

**Fig. 1.** Positions of allele-specific primers used for rapid cycle real-time PCR. (**A**) (+): Schematic presentation of the allele with 10 bp insertion, shown in red, (-): allele without insertion. Primers are shown as yellow arrows in 3′direction. PCR product lengths are 173 bp (+) and 163 bp (-) for primer pair H1/R3, 146 bp for the (+) allele-specific product for H1/R2 primer pair, and 138 bp for the (-) allele-specific product for H1/R1 primer pair. (**B**) Sequence context of the allele-specific priming on the *F-VIIc* promotor. The insertion sequence in the (+) allele is shown in red and is underlined

**Genotyping**

DNA samples of 91 healthy Caucasians were genotyped with the allele-specific real-time PCR assay. The results were confirmed by capillary electrophoresis. All samples showed identical genotyping results for both methods.

Among those tested, 59 were found to be homozygous for the deletion (-/-), one was homozygous for the 10-bp insertion (+/+), and 31 were heterozygous (+/-). These findings correspond to allele frequencies of 0.82 for the (-)-allele and 0.18 for the (+)-allele, agreeing with published data [8].

A genotyping experiment performed by allele-specific real-time PCR consists of two individual amplification reactions per DNA sample. One reaction contains the (+)-specific primers (H1/R2), the second reaction contains the (-)-specific primer pair (H1/R1). Each reaction contains 20 ng human genomic DNA as template. In the case of a matching primer pair, log-linear amplification is observed between cycles 24 to 30. For instance, a homozygous (+/+) sample results in an amplification curve with a $C_T$-value between 24 and 30 with the primers H1/R2. The amplification curve for H1/R1 (which is specific for the [-]-allele) has a $C_T$-value that is at least 10 cycles higher. For a heterozygous sample, the $C_T$-values of both amplification curves are identical. The (-)-specific primers do not amplify the (+)-template at all, resulting in an amplification curve only with the (-)-allele.

Late amplification of (+)-specific primers with (-)-alleles produces a product which is identical in both melting temperature and length to the product ampli-

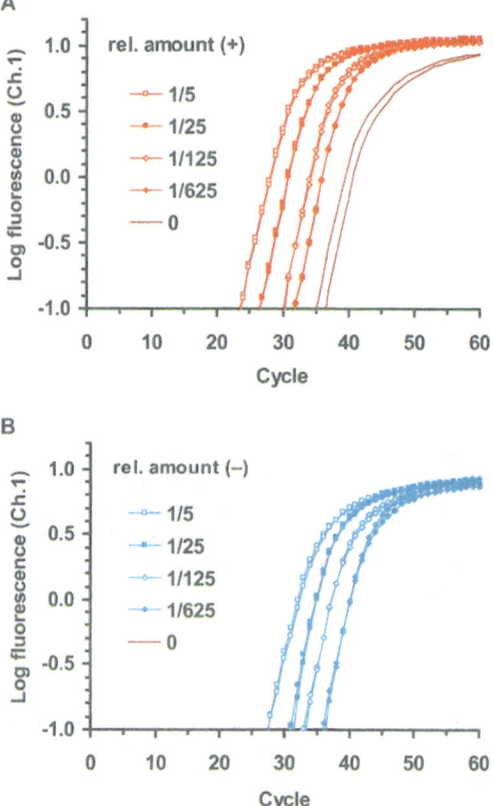

**Fig. 2.** Quantitative chimerism analysis for the *F-VIIc* 10-bp polymorphism. The experiment simulates different degrees of genetic chimerisms. For this purpose, genomic DNA of homozygous individuals (+/+ and -/-) was mixed in known ratios. These mixtures served as samples for the quantitative chimerism analysis. The samples contained a constant amount (6600 copies) of the allele in excess, and between 1320 and 10 copies of the allele specific for the primer pair used. (**A**) Amplification profiles for detection of the allele with insertion (H1/R2 primers). The negative control (0) contains 6600 copies of the allele without insertion as template. (**B**) Amplification profiles for detection of the allele without insertion (H1/R1 primer system). The negative control (0) contains 6600 copies of the allele with insertion as template

fied with perfect matching primers. This suggests that even five 3′-terminal mismatches do not completely hinder the polymerase from elongating the primer R2.

In addition to identification of a desired target allele, quantitative chimerism analysis is capable of quantifying one allele in the presence of an excess of non-target alleles.

    *Quantitative Chimerism Analysis*

The target allele of *F-VIIc* (either insertion or deletion) can be precisely quantified in the presence of a 3000-fold excess of non-target allele (Figure 2 and 3). This discrimination is about 150-fold more powerful than conventional real-time PCR techniques (1:20 [12, 13]). Even though false priming cannot be circumvented completely, this level of discrimination is the highest ever reached in quantita-

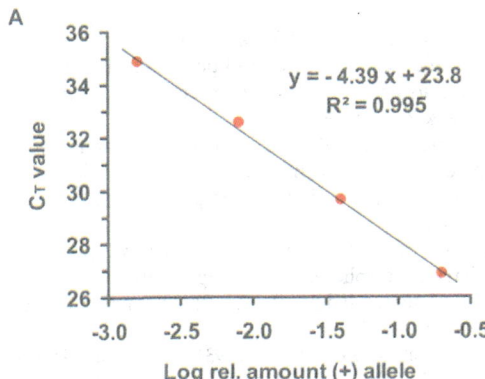

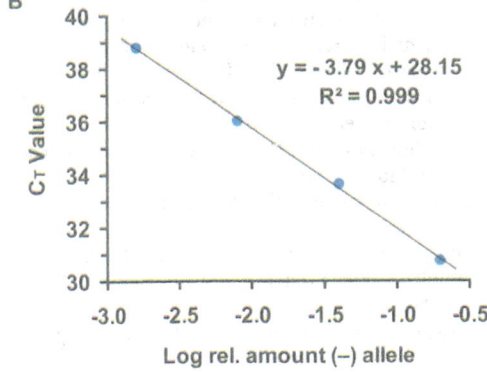

**Fig. 3.** Calibration plots for the amplification curves shown in Figure 2. The calibration plots demonstrate a high correlation between the $C_T$-value and the concentration of the amplified allele, despite the presence of high amounts of the mismatched allele. (**A**) Calibration plot for calculation of the insertion allele based on the data shown in Figure 2. (**B**) Calibration plot for calculation of the deletion allele. For calculation of the $C_T$-values the threshold was set to a value of 0.5 for both series of experiments

tive chimerism analysis [14] (except Y-chromosome specific techniques). The method developed here is inexpensive, fast, reliable, and easy to perform and, therefore, well suited for medical applications, *e.g.*, for monitoring graft *vs.* host disease in organ transplantation [15, 16, 17].

For a given sample with an unknown degree of chimerism, the degree of chimerism (D) can be estimated from the $\Delta C_T$-value of the two amplification curves by $D = 2^{\Delta C_T}$. The detailed derivation is given in [14].

## References

1. Fehse B, Chukhlovin A, Kuhlcke K, Marinetz O, Vorwig O, Renges H, Kruger W, Zabelina T, Dudina O, Finckenstein FG, Kroger N, Kabisch H, Hochhaus A, Zander AR (2001) Real-time quantitative Y chromosome-specific PCR (QYCS-PCR) for monitoring hematopoietic chimerism after sex-mismatched allogeneic stem cell transplantation. J Hematother Stem Cell Res 10:419–425

2. Formankova R, Honzatkova L, Moravcova J, Sieglova Z, Dvorakova R, Nadvornikova S, Vitek A, Lukasova M, Stary J, Brdicka R (2000) Prediction and reversion of post-transplant relapse in patients with chronic myeloid leukemia using mixed chimerism and residual disease detection and adoptive immunotherapy. Leuk Res 24:339–347

3. Sasanakul W, Hongeng S, Chuansumrit A, Chaiyaratana W, Pakakasama S, Hathirat P (2001) The usefulness of X-linked polymorphic loci as gene markers to track X allele and chimerism in a post-allogeneic peripheral blood stem cell transplant patient with Wiskott-Aldrich syndrome. J Med Assoc Thai 84:379–384

4. Jone CM, Akel N, Killeen AA (2000) Evaluation of chimerism in DNA samples by PCR amplification of D1S80 with detection by capillary electrophoresis. Mol Diagn 5:101–105

5. Wittwer CT, Fillmore GC, Hillyard DR (1989) Automated polymerase chain reaction in capillary tubes with hot air. Nucleic Acids Res 17:4353–4357

6. Marchetti G, Patracchini P, Papacchini M, Ferrati M, Bernardi F (1993) A polymorphism in the 5´region of coagulation factor VII gene (F7) caused by an inserted decanucleotide. Hum Genet 90:575–576

7. Di Castelnuovo A, D'Orazio A, Amore C, Falanga A, Donati MB, Iacoviello L (2000) The decanucleotide insertion/deletion polymorphism in the promoter region of the coagulation factor VII gene and the risk of familial myocardial infarction. Thromb Res 98:9–17

8. Humphries S, Temple A, Lane A, Green F, Cooper J, Miller G (1996) Low plasma levels of factor VIIc and antigen are more strongly associated with the 10 base pair promoter (-323) insertion than the glutamine 353 variant. Thromb Haemost 75:567–572

9. Wilhelm J, Pingoud A, Hahn M (2003) SoFAR: Software for fully automatic and highly accurate evaluation of real-time PCR data. BioTechniques 34:324–332

10. Wilhelm J, Pingoud A, Hahn M (2003) Validation of an algorithm for automatic quantification by real-time PCR. Anal Biochem 317:218–225

11. Hahn M, Wilhelm J, Pingoud A. (2001) Influence of fluorophor dye labels on the migration behavior of polymerase chain reaction–amplified short tandem repeats during denaturing capillary electrophoresis. Electrophoresis 22:2691–2700

12. Lay MJ, Wittwer CT (1997) Real-time fluorescence genotyping of factor V Leiden during rapid-cycle PCR. Clin Chem 43:2262–2267

13. Whitcombe D, Brownie J, Gillard HL, McKechnie D, Theaker J, Newton CR, Little S (1998) A homogeneous fluorescence assay for PCR amplicons: its application to real-time, single-tube genotyping. Clin Chem 44:918–923

14. Wilhelm J, Reuter H, Tews B, Pingoud A, Hahn M (2002) Detection and quantification of insertion/deletion-variations by allele-specific real-time-PCR: application for genotyping and chimerism analysis. Biol Chem 383:1423–1433

15. Kleeberger W, Rothamel T, Glockner S, Lehmann U, Kreipe H (2000) Laser-assisted microdissection and short tandem repeat PCR for the investigation of graft chimerism after solid organ transplantation. Pathobiol 68:196–201

16. Lagaaij EL, Cramer-Knijnenburg GF, van Kemenade FJ, van Es LA, Bruijn JA, van Krieken JH (2001) Endothelial cell chimerism after renal transplantation and vascular rejection. Lancet 357:33–37

17. Lion T (2001) Chimerism testing after allogeneic stem cell transplantation: Importance of timing and optimal technique for testing in different clinical-biological situations. Leukemia 15:292

# Applications

II

## Microbiology

# Detection of SARS-Coronavirus in the LightCycler by 5´-Nuclease Real-Time RT-PCR

CHRISTIAN DROSTEN*

## Introduction

The severe acute respiratory syndrome (SARS) is a recently discovered viral disease that involves an initial febrile phase followed by interstitial pneumonia and respiratory distress syndrome, leading to death in a fraction of patients. The case fatality proportion ranges from 13.2% in patients below 60 years of age and 43.3% in those above 60 (1). The first outbreak of SARS started in southern China in autumn 2002 and spread to several countries of the northern hemisphere until it came to rest in Juli 2003.

The causative agent of SARS has been identified to be a novel coronavirus (SARS-CoV) (2–4, 6). The virus seems to be highly contagious and environmentally stable.

Since antibodies are detectable only late in the disease (5), RT-PCR is the most promising method for diagnosing early acute cases. SARS-CoV can be detected best in respiratory samples and stool. Deep respiratory specimens, like sputum or endotracheal aspirates, contain the highest concentration of virus (2), but sampling goes along with the danger of producing infectious aerosols in the hospital. Swab specimens from the nasal cavity or the pharynx seem to contain less virus, but sampling is also a lot less dangerous. Stool samples are equally or more efficient than respiratory samples after day 10 of the disease (5), but they require special treatment to remove RT-PCR inhibitors.

From what is known by now, the sensitivity of RT-PCR is not sufficient to completely rule out the disease in patients presenting early after the onset of symptoms (5). Patients that match the clinical criteria of SARS should be repeatedly tested by RT-PCR after some days of symptoms, and suspects have to be formally ruled out after three or more weeks from the beginning of fever by serology.

This chapter describes two thoroughly evaluated real-time RT-PCR protocols for SARS-CoV. They can be used to achieve a valid confirmation of SARS in compliance with the guidelines issued by WHO during the 2003 SARS epidemic (refer to Text Box 1). Positive findings in either assay can be cross-confirmed by the other. Please note that these guidelines will be subject to change in the future to meet the requirements of non-epidemic settings.

* Christian Drosten, Bernhard-Nocht Institute for Tropical Medicine, Hamburg, Germany
  E-mail: drosten@bni-hamburg.de

**Text Box 1**

## Materials

**Equipment**

LightCycler Instrument (Roche Diagnostics, Mannheim, Germany)
LightCycler Reaction Capillaries (Roche Diagnostics, Mannheim, Germany)
LightCycler software version 3.5 (Roche Diagnostics, Mannheim, Germany)
Primer Express (Applied Biosystems, Weiterstadt, Germany)

**Reagents**

Amplification primers (Tib-Molbiol, Berlin, Germany)
Hybridization probes (Tib-Molbiol, Berlin, Germany)
Superscript II reverse transcriptase/Platinum Taq polymerase one-step RT-PCR kit (Invitrogen, Karlsruhe, Germany)
Megascript T7 Kit (Ambion, Austin, Texas)
pCRII Topo TA Cloning Kit (Invitrogen, Karlsruhe, Germany)

**Text Box 2**

## Procedure

**Preparation of Template DNA**

Template RNA was prepared in a biosafety level 3 laboratory as described in Table 1.

RNA solutions were stored at 8°C until further processing. 5 µl portions were lateron analyzed in PCR.

**Oligonucleotides and Reagent Formulations Protocol A**

Protocol A has originally been published in connection with the identification of the causative agent (2). It targets the replicase gene in open reading frame 1b of SARS-CoV, the only genomic region of SARS-CoV known at that point of time.

**Table 1.** Preparation of clinical specimens for RT-PCR detection of SARS-coronavirus

| Sputum and endotracheal aspirates | Swab specimen | Stool specimens |
|---|---|---|
| Use native sputum or endotracheal aspirate specimens<br>Mix with equal volume of 2 X NACC buffer[a]<br>Slowly shake for 30 min | Use native swabs containing no preservatives or trans-portmedia (e.g., Q-Tips whetted with one drop of 0.9% NaCl) | Use 200 mg or 200 µl native stool specimens |
| Add 140 µl of homogenate to 560 µl of buffer AVL (Viral RNA mini kit, Qiagen) Proceed according to manu-facturer's instructions | Dip whole swab into 560 µl of buffer AVL (Viral RNA mini kit, Qiagen)<br>Add 500 µl Ethanol abs.<br>Proceed according to manu-facturer′s instructions | Treat samples exactly accord-ing to manufacturer's instruc-tions, Qiagen DNA stool mini kit, protocol for isolation of DNA from stool for pathogen detection |
| Elute RNA from column with 60 µl 80°C elution buffer (AVE, included in the kit) | | Elute RNA from column with 200 µl 80°C elution buffer (AE, included in the kit) |

[a]1% N-acetylcysteine, 0.9% NaCl, in double-destilled water

Like in other RNA viruses, the polymerase/replicase gene of SARS-CoV is high-ly conserved between different isolates (8) and can thus be considered a reliable RT-PCR target. The assay uses a 5′-nuclease probe, a detection format chosen in the intention of extending the compatibility of the protocol to real-time PCR instruments other than the LightCycler. The assay described in protocol A is the basis of the first commercial RT-PCR kit for the virus. The analytical sensitivity of the protocol is the same as that of the kit, but the kit in addition contains an internal control.

**Protocol A.** Replicase gene RT-PCR assay for detection of SARS-coronavirus

| Oligonucleotides replicase gene assay | | | |
|---|---|---|---|
| Name<br>Function | Genome position<br>SARS-CoV Urbani<br>Acc. AY278741 | Sequence<br>5′ → 3′ | $T_m$ (°C) |
| BNITMSARS1<br>Forward primer | 18187–18206 | TTATCACCCGCGAAGAAGCT | 63.5 |
| BNITMSARAS2<br>Reverse primer | 18243–18264 | CTCTAGTTGCATGACAGCCCTC | 62.7 |
| TMSARP1<br>Probe | 18218–18241 | FAM-TCGTGCGTGGATTGGCTT-TGATGT-TAMRA | 73.0 |

| Formulation replicase gene assay | | |
|---|---|---|
| Component | Stock concentration | Volume per reaction |
| Reaction mix (Superscript II/ Platinum Kit, Invitrogen) | 2.5 X | 12.5 µl |
| Non-acetylated BSA | 1mg/ml | 1 µl |
| $MgSO_4$ | 50mM | 1.2 µl |
| BNITMSARS1 | 10 µM | 0.5 µl |
| BNITMSARAS2 | 10 µM | 0.5 µl |
| BNITMSARP | 10 µM | 0.6 µl |
| RT/*Taq* Mixture (Superscript II/ Platinum Kit, Invitrogen) | as provided | 0.6 µl |
| $H_2O$ (PCR grade) | – | 3.1 µl |
| RNA solution | – | 5 µl |

**Protocol B**

Protocol B targets the nucleocapsid (N) gene of SARS-CoV and uses the same reagents as protocol A. It can be used to achieve a valid confirmation of SARS in compliance with the guidelines issued by WHO during the 2003 SARS epidemic (refer to Text Box 1). Use of the N gene was based on the reportedly higher abundance of such subgenomic RNA in infected cells, possibly leading to an increased detection sensitivity. However, in a preliminary study described below, it appeared that the assay is not more sensitive than the one described in Protocol A. Protocol B uses 5´-nuclease probes for the same reasons as stated above.

**Protocol B.** Nucleoprotein gene RT-PCR assay for detection of SARS coronavirus

| Oligonucleotides nucleoprotein gene assay | | | |
|---|---|---|---|
| Name Function | Genome position SARS-CoV Urbani Acc. AY278741 | Sequence 5´ → 3´ | $T_m$ (°C) |
| SANS1 Forward primer | 28176–28196 | TGGACCCACAGATTCAACTGA | 62.7 |
| SANAs2 Reverse primer | 28265–28286 | GCTGTGAACCAAGACGCAGTAT | 63.0 |
| SANP Probe | 28200–28223 | FAM-TAACCAGAATGGAGGACG-CAATGGT(TAMRA-T)P | 69.6 |

| Formulation nucleoprotein gene assay | | |
| --- | --- | --- |
| Component | Stock concentration | Volume per reaction |
| Reaction mix (Superscript II/ Platinum Kit, Invitrogen) | 2.5 X | 12.5 µl |
| Non-acetylated BSA | 1mg/ml | 1 µl |
| MgSO$_4$ | 50mM | 1.2 µl |
| SANS1 | 10 µM | 0.5 µl |
| SANAs2 | 10 µM | 1 µl |
| SANP | 10 µM | 0.5 µl |
| RT/*Taq* Mixture (Superscript II/ Platinum Kit, Invitrogen) | as provided | 0.5 µl |
| H$_2$O (PCR grade) | – | 2.8 µl |
| RNA solution | – | 5 µl |

Master mixes of the described reaction mixtures were freshly prepared, containing all reaction components except the 5 µl of RNA solution. The master mix was then distributed in volumes of 20 µl into LightCycler reaction capillaries, 5 µl of RNA solution were added, and the content of each capillary was spun down by a brief centrifugation step after capping. For each patient RNA solution to be tested, a parallel reaction was prepared, containing the same 5 µl of RNA solution and the same 20 µl of reaction mix. To allow detection of substances interfering with PCR amplification (PCR inhibitors), an additional 1 µl of an RNA run control solution was added. The run control solution had previously been prepared by end-point dilution of a high-titered RNA extract. The concentration of the run-control solution was adjusted to range 1.5 Log10 above the detection end point of that dilution series. The resulting solution was then stored in small aliquots at –20°C for subsequent use.

Capillaries were placed in a LightCycler Carousel and subjected to thermal cycling as follows.

**Amplification and Inhibition Control**

**Protocol A**    Thermal cycling profile for replicase gene RT-PCR assay (Protocol A)

| Parameter | Value | |
|---|---|---|
| Reverse transcription | | |
| Cycles | 1 | |
| | Segment 1 | |
| Target temperature [°C] | 45°C | |
| Incubation time [min] | 20 | |
| Temperature transition rate [°C/s] | 20 | |
| Acquisition mode | none | |
| Gains | Automatic | |
| | | |
| Denaturation | | |
| Cycles | 1 | |
| | Segment 1 | |
| Target temperature [°C] | 95°C | |
| Incubation time [min] | 3 | |
| Temperature transition rate [°C/s] | 20 | |
| Acquisition mode | none | |
| Gains | Automatic | |
| | | |
| Amplification | | |
| Cycles | 40 | |
| | Segment 1 | Segment 2 |
| Target temperature [°C] | 95 | 58 |
| Incubation time [s] | 10 | 30 |
| Temperature transition rate [°C/s] | 20 | 20 |
| Acquisition mode | None | Single |
| Gains | Automatic | |

Thermal cycling profile for nucleocapsid gene RT-PCR assay (Protocol B)

| Parameter | Value | | |
|---|---|---|---|
| Reverse transcription | | | |
| Cycles | 1 | | |
| Segment 1 | | | |
| Target temperature [°C] | 50°C | | |
| Incubation time [min] | 10 | | |
| Temperature transition rate [°C/s] | 20 | | |
| Acquisition mode | none | | |
| Gains | Automatic | | |
| | | | |
| Denaturation | | | |
| Cycles | 1 | | |
| | Segment 1 | | |
| Target temperature [°C] | 95°C | | |
| Incubation time [min] | 3 | | |
| Temperature transition rate [°C/s] | 20 | | |
| Acquisition mode | none | | |
| Gains | Automatic | | |
| | | | |
| Amplification | | | |
| Cycles | 40 | | |
| | Segment 1 | Segment 2 | Segment 3 |
| Target temperature [°C] | 95 | 55 | 72 |
| Incubation time [s] | 2 | 12 | 10 |
| Temperature transition rate [°C/s] | 20 | 20 | 20 |
| Acquisition mode | None | Single | None |
| Gains | Automatic | | |

Fluorescence in both protocols was read on the F1/F2 channel. A reaction was considered positive when an exponential increase in fluorescent signal was observed, negative when there was no signal in the patient test reaction but a signal in the corresponding inhibition control reaction (spiked with run control solution), and inhibited when there was no signal in the test reaction and the inhibition control reaction. Refer to Text Box 1 for diagnostic implications of results.

Both protocols allowed to quantify the virus RNA concentration (note that the term "viral load" should be reserved for HIV-1 plasma vireamia). To serve as reference standards for real-time RT-PCR quantification, in-vitro transcribed RNA was generated for both target genes by cloning PCR products of each assay into a plasmid vector with the pCRII Topo TA Cloning Kit (Invitrogen, Karlsruhe, Germany), which was then transcribed into RNA by T7 polymerase (Megascript T7

Kit, Ambion, Austin, Texas). The RNA solutions were photometrically quantified and diluted in double destilled DEPC-treated water containing 25 μg/ml *E. coli* tRNA as a non-specific carrier. Aliquots were stored at −80°C. Dilution series ranging from concentrations of about 10 to 10.000 copies per 140 μl were extracted with the Qiagen viral RNA kit and amplified in parallel with the experimental samples to be quantified. The concentrations of the standard solutions were entered in the LightCycler sample sheet. Evaluation then followed the guidelines provided in the LightCycler user´s manual for absolute quantification of target genes.

## Results

The analytical sensitivities of both protocols were determined using in-vitro transcribed RNA standards, generated by cloning PCR products of each assay into a plasmid vector that was then transcribed into RNA by T7 polymerase. Replicate reactions containing different amounts of RNA transcripts (2, 4, 8, and 16 copies of RNA per reaction) were conducted, and the results were submitted to probit regression analysis. The replicase and the N assay, respectively, yielded positive results at a 50% chance with a theroretical 1.9 or 2.2 copies of RNA per assay, and at a 95% chance with 2.8 or 3.0 copies per assay. It could therefore be assumed that the tests have equivalent analytical sensitivities.

In clinical samples, an N gene based test would nevertheless be expected to yield a better sensitivity, since subgenomic N fragments have been reported to be highly abundant in cells (7). To test this hypothesis, quantitative RT-PCR results from both protocols were compared in a total of 21 samples from 5 patients with SARS. Samples and history of patients A and B have earlier been described in detail (2). Patient C was a male German who had returned to Germany from Hanoi, Vietnam after a business trip. Patients D and E were hospitalized in Hong Kong (samples kindly provided by Dr. John Tam, Chinese University of Hong Kong). Seroconversion in all patients had been demonstrated after convalescence. Quantification was done by co-amplifying a limiting dilution series of in-vitro transcribed RNA standards. The RNA copy number in all samples was equivalent in both assays (Table 2), suggesting that the N gene assay would not increase the clinical sensitivity in larger patient cohorts.

Figure 1 shows that the RNA quantification results in both assays correlate perfectly (r=1.0).

The replication kinetics of SARS-CoV gives an explanation for these somewhat unexpected findings (Figure 2). The cytoplasma of Vero cells, infected with SARS-CoV at a multiplicity of infection of 0.01, indeed yielded about 5 times more N gene RNA than replicase gene RNA on the first day of infection. After a few days, however, levels of both RNAs approximated each other. In the supernatant, abundance of both RNAs was equivalent throughout the infection. If one assumes that infected cells contained in clinical samples are already in a late stage of virus replication, it is therefore no surprise that both target genes yield the same diagnostic outcome in RT-PCR.

**Table 2.** Quantitative detection of SARS-CoV in clinical samples

| Patient Type of sample | Virus RNA concentration, copies per sample | |
|---|---|---|
| | Replicase assay | Nucleocapsid assay |
| **A** | | |
| Nasal swab | $1.2 \times 10^2$ | $2.5 \times 10^1$ |
| Pharyngeal swab | $8.6 \times 10^2$ | $8.5 \times 10^2$ |
| Sputum[a] | $5.5 \times 10^6$ | $6.1 \times 10^6$ |
| Plasma | $5.5 \times 10^2$ | $7.8 \times 10^2$ |
| Stool[b] | $1.6 \times 10^3$ | $1.6 \times 10^3$ |
| Broncho-alveolar lavage | $1.7 \times 10^5$ | $2.9 \times 10^5$ |
| Pharyngeal swab, late[b] | – | – |
| **B** | | |
| Nasal swab | $7.4 \times 10^1$ | $9.8 \times 10^1$ |
| Pharyngeal swab | – | – |
| Sputum[a] | $1.9 \times 10^4$ | $2.4 \times 10^4$ |
| Plasma | – | – |
| Stool[c] | $6.1 \times 10^4$ | $6.9 \times 10^4$ |
| Pharyngeal swab, late[c] | – | – |
| **C** | | |
| Pharyngeal swab | – | – |
| Serum | – | – |
| Stool[d] | $3.2 \times 10^1$ | $4.4 \times 10^1$ |
| Broncho-alveolar lavage | $3.0 \times 10^6$ | $3.5 \times 10^6$ |
| Pharyngeal swab, late[d] | – | – |
| Urine[d] | – | – |
| **D** | | |
| Pharyngeal swab | $7.4 \times 10^2$ | $1.1 \times 10^3$ |
| **E** | | |
| Pharyngeal swab | $1.1 \times 10^6$ | $1.4 \times 10^6$ |

Patient A. Samples from day 9, unless otherwise stated
Patient B. Samples from day 3, unless otherwise stated
Patient C. Samples from day 4, unless otherwise stated
Patient D. Unknown sampling date
Patient E. Unknown sampling date
a. Specimen prediluted due to limited availability of material
b. After convalescence, day 25
c. After convalescence, day 19
d. After convalescence, day 17

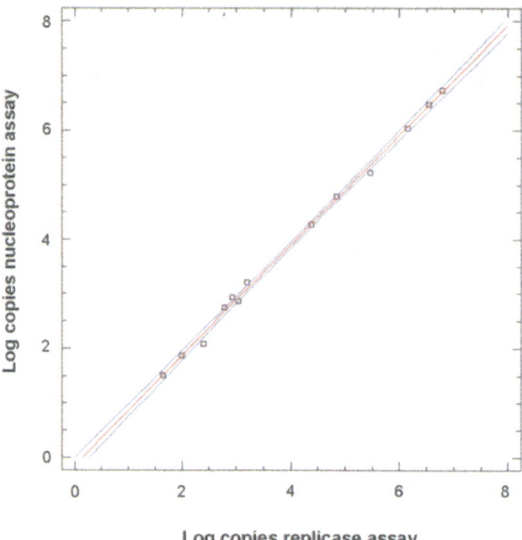

**Fig. 1.** Virus concentration in 21 Samples from 5 patients with confirmed SARS (refer to Table 2) as determined with replicase and nucleoprotein gene real-time RT-PCR assays. Black squares represent data points, the red line is a linear regression line, the blue lines are 95% confidence limits of the regression. Statistical evaluation was done with the Statgraphics 5 package, Statistical Graphics, Inc

## Comments

**General**

The absence of an 100% sensitive laboratory test for SARS, especially in the early phase of the disease, will become a major problem in case management during influenza epidemics. One infectious patient can start a SARS epidemic, but SARS is a very unlikely diagnosis at all. The protocols presented in this chapter enable a fully valid laboratory confirmation of SARS cases by real-time RT-PCR. However, the known pitfalls of PCR diagnostics, like contamination or interpretation errors, are especially critical in this disease; a false positive result can cause tremendous public disconcertment that may have unforseeable adverse effects. In a situation of a low (zero) level of SARS endemicity with other similar diseases prevailing, suspected cases must therefore be handled with utmost care. Positive results should in any case be confirmed by a SARS reference laboratory. Contact WHO for an up-to-date listing of such insitutions.

**Choice of Sample Type**

It is essential to notice that according to experiences with SARS patients in Singapore (Evelyn Koay, National University of Singapore, personal communication) and Germany the virus RNA concentration in clinical specimens declines in the following order: sputum or endotracheal aspirates>stool samples>pharyngeal swabs or saliva. Sampling of sputum and endotracheal aspirates generates infec-

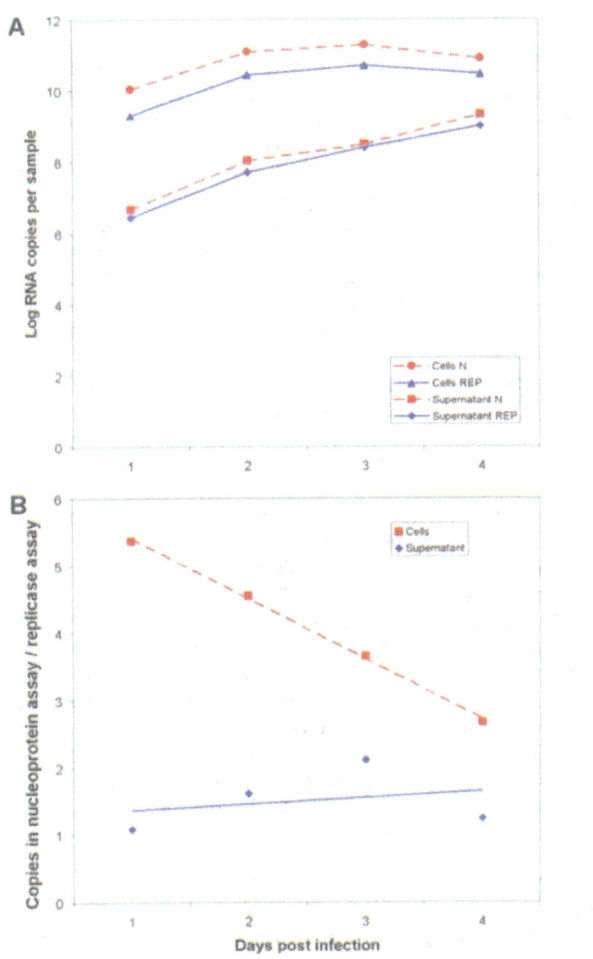

**Fig. 2.** Accumulation kinetics of SARS-Coronavirus replicase and nucleoprotein genes in Vero cell culture. **A.** Intracellular copy numbers ("Cells") versus copy numbers in supernatant. REP: replicase gene, N: nucleoprotein gene. **B.** Copy numbers of nucleoprotein gene per copy numbers of replicase gene in cells and supernatants

tious aerosols and can therefore be done only in negative pressure isolation units with adequate respiratory protection for health care workers. Stool samples are more appropriate in the later phase of the disease (beyond day 10, (5)).

SARS-CoV is highly contagious, and laboratory-associated infections have already occurred in the past. Note that according to WHO guidelines, biosafety level 2 is required for handling clinical specimens prior to PCR. Manipulations such as virus culture require at least biosafety level 3.

**Laboratory Safety Requirements**

# References

1. Donnelly, C. A., A. C. Ghani, G. M. Leung, A. J. Hedley, C. Fraser, S. Riley, L. J. Abu-Raddad, L. M. Ho, T. Q. Thach, P. Chau, K. P. Chan, T. H. Lam, L. Y. Tse, T. Tsang, S. H. Liu, J. H. Kong, E. M. Lau, N. M. Ferguson, and R. M. Anderson. 2003. Epidemiological determinants of spread of causal agent of severe acute respiratory syndrome in Hong Kong. Lancet. 361:1761–6

2. Drosten, C., S. Gunther, W. Preiser, S. van der Werf, H. R. Brodt, S. Becker, H. Rabenau, M. Panning, L. Kolesnikova, R. A. Fouchier, A. Berger, A. M. Burguiere, J. Cinatl, M. Eickmann, N. Escriou, K. Grywna, S. Kramme, J. C. Manuguerra, S. Muller, V. Rickerts, M. Sturmer, S. Vieth, H. D. Klenk, A. D. Osterhaus, H. Schmitz, and H. W. Doerr. 2003. Identification of a novel coronavirus in patients with severe acute respiratory syndrome. N Engl J Med. 348:1967–76

3. Fouchier, R. A., T. Kuiken, M. Schutten, G. van Amerongen, G. J. van Doornum, B. G. van den Hoogen, M. Peiris, W. Lim, K. Stohr, and A. D. Osterhaus. 2003. Aetiology: Koch's postulates fulfilled for SARS virus. Nature. 423:240

4. Ksiazek, T. G., D. Erdman, C. S. Goldsmith, S. R. Zaki, T. Peret, S. Emery, S. Tong, C. Urbani, J. A. Comer, W. Lim, P. E. Rollin, S. F. Dowell, A. E. Ling, C. D. Humphrey, W. J. Shieh, J. Guarner, C. D. Paddock, P. Rota, B. Fields, J. DeRisi, J. Y. Yang, N. Cox, J. M. Hughes, J. W. LeDuc, W. J. Bellini, and L. J. Anderson. 2003. A novel coronavirus associated with severe acute respiratory syndrome. N Engl J Med. 348:1953–66

5. Peiris, J. S., C. M. Chu, V. C. Cheng, K. S. Chan, I. F. Hung, L. L. Poon, K. I. Law, B. S. Tang, T. Y. Hon, C. S. Chan, K. H. Chan, J. S. Ng, B. J. Zheng, W. L. Ng, R. W. Lai, Y. Guan, and K. Y. Yuen. 2003. Clinical progression and viral load in a community outbreak of coronavirus-associated SARS pneumonia: a prospective study. Lancet. 361:1767–72

6. Peiris, J. S., S. T. Lai, L. L. Poon, Y. Guan, L. Y. Yam, W. Lim, J. Nicholls, W. K. Yee, W. W. Yan, M. T. Cheung, V. C. Cheng, K. H. Chan, D. N. Tsang, R. W. Yung, T. K. Ng, and K. Y. Yuen. 2003. Coronavirus as a possible cause of severe acute respiratory syndrome. Lancet. 361:1319–25

7. Rota, P. A., M. S. Oberste, S. S. Monroe, W. A. Nix, R. Campagnoli, J. P. Icenogle, S. Penaranda, B. Bankamp, K. Maher, M. H. Chen, S. Tong, A. Tamin, L. Lowe, M. Frace, J. L. DeRisi, Q. Chen, D. Wang, D. D. Erdman, T. C. Peret, C. Burns, T. G. Ksiazek, P. E. Rollin, A. Sanchez, S. Liffick, B. Holloway, J. Limor, K. McCaustland, M. Olsen-Rasmussen, R. Fouchier, S. Gunther, A. D. Osterhaus, C. Drosten, M. A. Pallansch, L. J. Anderson, and W. J. Bellini. 2003. Characterization of a novel coronavirus associated with severe acute respiratory syndrome. Science. 300:1394–9

8. Ruan, Y. J., C. L. Wei, A. L. Ee, V. B. Vega, H. Thoreau, S. T. Su, J. M. Chia, P. Ng, K. P. Chiu, L. Lim, T. Zhang, C. K. Peng, E. O. Lin, N. M. Lee, S. L. Yee, L. F. Ng, R. E. Chee, L. W. Stanton, P. M. Long, and E. T. Liu. 2003. Comparative full-length genome sequence analysis of 14 SARS coronavirus isolates and common mutations associated with putative origins of infection. Lancet. 361:1779–85

# Normalized Quantitative Rapid-Cycle Real-Time PCR for the Assessment of CMV-DNA Copies

Markus Stöcher and Jörg Berg*

## Introduction

Molecular assays based on real-time PCR are widely used for the detection and quantification of virus-derived nucleic acids in clinical specimens (1). When quantification is performed with real-time PCR, sample nucleic acids are commonly tested against reference nucleic acids used as external quantification standards. The external quantification standards are applied in graded amounts and assayed in parallel to sample nucleic acids. This quantification approach assumes that equal amplification efficiencies exist in quantification standards and in samples. However, this prerequisite may not always be met. Clinical samples contain PCR inhibiting moieties such as hemoglobin, myoglobin or immunoglobins (2). These moieties may not always reliably or completely be eliminated in the process of nucleic acid purification. Therefore, remnants hereof may hamper the amplification reaction. So, amplifications may vary from sample to sample including those containing quantification standards.

In order to monitor and to normalize for potential varying amplification efficiencies, internal amplification controls comprised of heterologous or homologous DNA fragments have been used in conventional quantitative PCR assays, to assess copy numbers of virus-specific nucleic acids (3–5).

In the present contribution we describe the development of a quantitative competitive real-time PCR approach on the LightCycler instrument using co-amplification of an internal control (IC) to normalize possible varying amplifications between standard samples and test samples prior to quantification calculation. We call this approach normalized quantitative competitive LightCycler PCR (NQC-LC-PCR) and have applied this methodology to the quantification of CMV DNA in plasma samples (6).

## Materials

LightCycler™ Instrument (Roche Applied Science, Mannheim, Germany)    Equipment
MagNA Pure LC™ Instrument (Roche Applied Science, Mannheim, Germany)
LightCycler™ carousel centrifuge (Roche Applied Science, Mannheim, Germany)

* Jörg Berg, Institute of Laboratory Medicine, General Hospital Linz, Krankenhausstr. 9, A-4020 Linz, Austria, e-mail: Joerg.Berg@akh.linz.at

Conventional thermocycler, e.g. Touchdown (Thermo Hybaid Ltd., Ashford, UK);
Incubator (37°C), e.g. Heraeus-Kendro, (Heraeus-Kendro, Hanau, Germany)
Shaking water-bath, e.g. GFL1092 (Gesellschaft für Labortechnik mbH, Burgwedel, Germany)
Agarose-gel electrophoresis unit, e.g. Sub-Cell GT (Biorad, Richmond, CA, USA)
Spectrophotometer, e.g. U-2001 (Hitachi Instrument INC, USA)
Centrifuge, e.g. Multifuge 3 S-R (Heraeus-Kendro, Hanau, Germany)
Microcentrifuge, e.g. Biofresco (Heraeus-Kendro, Hanau, Germany)

**Reagents**

Primers (TIB MOLBIOL, Berlin, Germany and MWG-Biotech, Munich, Germany)
Probes (TIB MOLBIOL, Berlin, Germany)
High Pure Viral Nucleic Acid Kit (Roche Applied Science, Mannheim, Germany)
LightCycler™ FastStart DNA Master Hybridization Probes (Roche Applied Science, Mannheim, Germany)
Purified quantitated genomic CMV DNA (strain AD 169) (Advanced Biotechnologies, Columbia, MD, USA)
Taq-polymerase and PCR reagents, e.g. AmpliTaq gold (Applied Sciences, Fostercity, CA, USA)
TA-cloning kit, e.g. AdvanTage™ PCR cloning kit (Clontech Laboratories Inc., Palo Alto, CA, USA)
Plasmid DNA extraction Kit, e.g. Quantum Plasmid Miniprep Kit (Biorad, Richmond, CA, USA)
Restriction enzyme Hind III (New England Biolabs Inc., Beverly, MA, USA);

## Procedure

**Study Design**

A rapid-cycle real-time PCR specific for the detection CMV-DNA was established using purified genomic CMV-DNA as reference standard. An internal amplification control (IC) was generated as heterologous PCR competitor. The amplification efficiencies of the CMV specific amplification and the IC-specific amplification were assessed. Calibrator samples were prepared as run controls, which contain equal amounts of reference CMV DNA and IC DNA. An algorithm was developed to normalize amplification variations by comparing the IC amplification in the calibrator sample with those of the test samples. Normalized amplification values were introduced into a further algorithm to quantify CMV-DNA copies in test samples, which were tested against known CMV-DNA copies in the calibrator sample. This normalized quantitative competitive LightCycler PCR (NQC-LC-PCR) was initially set up with a calibrator sample that consisted of about 100 copies of CMV reference standard DNA and about 100 copies of IC DNA. Then, the NQC-LC-PCR approach was compared with quantification using an external CMV standard in graded amounts. The power of the normalization of NQC-LC-PCR was studied with test samples containing PCR inhibitors in graded amounts.

A CMV specific assay with the NQC-LC-PCR methodology was established by combining CMV-specific NQC-LC-PCR with DNA purification on the MagnaPure

LC instrument. For this assay the calibrator sample contained 5000 copies/ml of CMV reference standard and of 5000 copies/ml of IC DNA. During the automated nucleic acid purification the IC DNA was introduced into each test sample to yield the same concentration as in the calibrator sample. Detection limits and range of quantification were determined by spiking the reference standard DNA in graded amounts into normal plasma together with a defined amount of internal control DNA. For the evaluation of the novel assay clinical specimens were tested, and quantification results were compared with quantification by external quantification standards.

**Sample Preparation**

DNA from normal plasma spiked with CMV reference standard DNA and DNA from clinical samples were purified on the MagNA Pure LC instrument (Roche Applied Science, Vienna, Austria). The MagNA Pure LC Total Nucleic Acid Isolation Kit-large volume (Roche Applied Science, Vienna, Austria) was used and sample volume of 1 ml was chosen. Elution was performed with 50 µl of elution buffer.

Calibrator samples and CMV external standard samples were prepared by combining 400 µl of the MagNA Pure LC Total Nucleic Acid Isolation Kit-large volume lysis buffer with 595 µl normal plasma and with 5 µl of serially diluted CMV reference standard DNA. IC DNA was added to the MagNA Pure LC Total Nucleic Acid Isolation Kit-large volume lysis buffer of each automated purification run to yield approximately 5000 copies/ml sample.

**Primers and Probes**

The sequences and the characteristics of the primers and probes are out-lined in Table 1. The primers for the preparative PCR of the IC were composed of the CMV specific primer sequences each in 5' of the primer sequences that target a stretch of the neomycin phosphotransferase gene (neo) used as heterologous DNA.

The CMV-specific primers target the US17 gene of the CMV genome and were previously used in diagnostic applications to the detection of CMV-DNA in clinical samples (7). A pair of CMV-specific FRET hybridization probes were labeled with Fluorescein and LC-Red 705 and used to detect CMV-specific PCR products. The IC specific products were detected with a pair of neo specific FRET hybridization probes that were labeled with Fluorescein and LC-Red 640.

**Generation of Internal Control**

The CMV-specific IC was devised as PCR competitor. It consisted of the CMV-specific forward and reverse primer sequences, which flanked a stretch of heterologous DNA. As heterologous DNA the bacterial neomycin phosphotransferase gene (neo) was used (8). As recently described, preparative PCR was carried out in a conventional thermocycler using the composite primers (see table 1) and a plasmid containing neo (8). This generated a 325 bp fragment, which was subsequently cloned into the plasmid vector pT-Adv according to the manufacturer's instruction. The plasmid DNA was purified from transformed bacteria by a miniprep procedure and linearized by Hind III restriction. The DNA concentration was assessed by UV spectrophotometry at λ 260 nm with reference at λ 280 nm.

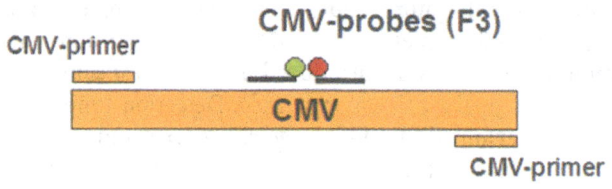

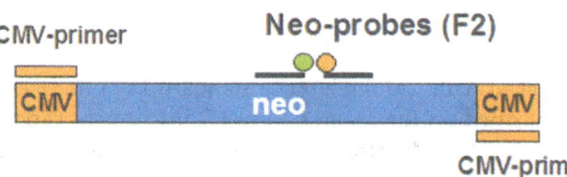

**Fig. 1.** Schematic representation of the CMV-specific and IC-specific PCR products, which were detected by their specific FRET hybridization probes. The CMV-specific 3' FRET-probe was labeled with LC-Red705 and detected on Channel F3, whereas the IC-specific 3' FRET-probe was labeled with LC-Red640 and detected on channel F2 of the LightCycler instrument

**Table 1.** Oligonucleotides

| CMV (GenBank Accession # NC 001347) | | | |
|---|---|---|---|
| | Position | Length (bp) | $T_m$ (°C) |
| **Primers** | | | |
| GAC ACA ACA CCG TAA AGC | 120,209 | 18 | 60.2 |
| CAG CGT TCG TGT TTC C | 120,487R | 16 | 59.7 |
| **Hybridization probes** | | | |
| TTG CGG GTC ATC GTC AGG TCC TC-F | 120,262 | 23 | 71.7 |
| LCRed705-TCC ACG TCA GAG CCC AGC GTG A | 120,286 | 22 | 73.5 |
| **NEO (GenBank Accession # M28248)** | | | |
| **Composite Primers for the Generation of the IC in preparative PCR** | | | |
| | Position | Length (bp) | $T_m$ (°C) |
| GAC ACA ACA CCG TAA AGC - CGG TGC CCT GAA TGA ACT[a] | 2226[b] | 18 + 18 | 61.9[c] |
| CAG CGT TCG TGT TTC C - ACC GGC TTC CAT CCG A[a] | 2519R[b] | 16 + 16 | 62.3[c] |
| **Hybridization probes for the detection of IC in LightCycler-PCR** | | | |
| GCT GCA TAC GCT TGA TCC GGC T -F | 516 | 22 | 71.1 |
| LCRed640-CCT GCC CAT CGA CCA CCA AG C | 539 | 22 | 71.6 |

[a] neo-specific sequences are underlined
[b] positions of the neo-specific stretch of the composite primers on the neo gene
[c] Tm exclusively related to neo specific amplification

The master mixtures for each 20 µl reactions were prepared as follows:

|  | Volume [µl] | [Final] |
|---|---|---|
| LightCycler™ FastStart DNA Master Hybridization Probes | 2.0 | 1x |
| MgCl₂ (25 mM) | 2.4 | 4 mM |
| Primers (20 µM/ml each) | 0.4 + 0.4 | 0.4µM each |
| CMV-Probes (20 µM/ml each) | 0.2 + 0.2 | 0.2µM each |
| Neo-Probes (20 µM/ml each) | 0.2 + 0.2 | 0.2µM each |
| H₂O (PCR grade) | 9.0 |  |
| Total Volume | 15.0 |  |

The capillaries were loaded with 15 µl of the reaction mixture and with 5 µl of sample DNA and then placed into the carousel of the LightCycler instrument. The carousel was spun down in the dedicated centrifuge at 735 x g for 15 s and placed into the LightCycler instrument. A color compensation file was generated according to the manufacturer's instruction and applied to all PCR runs. All runs were accompanied by a negative control sample containing IC, a calibrator sample and a reagent control containing PCR-grade water instead of DNA.

The following cycling program was used.
- Denaturation at 95°C for 7 min
- Amplification

| Parameter | Value | | |
|---|---|---|---|
| Cycles | 50 | | |
| Type | Quantification | | |
| | Segment 1 | Segment 2 | Segment 3 |
| Target temperature [°C] | 95 | 60 | 72 |
| Incubation time [s] | 2 | 10 | 15 |
| Temperature transition rate [°C/s] | 20 | 20 | 20 |
| Acquisition mode | none | single | none |
| Gains | automatic | | |

- Melting Curve Analysis:

| Parameter | Value | | | |
|---|---|---|---|---|
| Cycles | 1 | | | |
| Type | Melting Curve | | | |
| | Segment 1 | Segment 2 | Segment 3 | Segment 4 |
| Target temperature [°C] | 95 | 60 | 50 | 85 |
| Incubation time [s] | 60 | 120 | 2 | |
| Temperature transition rate [°C/s] | 20 | 20 | 20 | 0.2 |
| Acquisition mode | none | none | none | continuous |
| Gains | automatic | | | |

- Cooling to 40°C for 30 s.

CMV and IC specific products were detected on channels F3 and F2, respectively. The CMV and IC specific melting temperatures were 72°C and 70°C, respectively. Crossing point (CP) values were determined by applying the fit points algorithm with two points (LightCycler software version 3.5.3).

**Normalization**

In a first step the efficiencies of the CMV-specific and the IC-specific amplifications were determined. Standard curves for both amplification reactions were generated using the CMV reference standard DNA serially diluted and spiked with IC DNA and vice versa. Thereafter, logarithmic linear regression analysis was performed by the LightCycler instrument (LightCycler software 3.5.3), which resulted in the slope values $S_{CMV}$, (CMV-specific amplification) and $S_{IC}$ (IC-specific amplification) according to following equations:

$$CP_{CMV} = S_{CMV} \times \log C_{CMV} + Y_{CMV} \tag{1}$$

$$CP_{IC} = S_{IC} \times \log C_{IC} + Y_{IC} \tag{2}$$

$CP_{CMV}$ and $CP_{IC}$ correspond to the CP values of the CMV-specific and IC-specific amplifications, respectively; $C_{CMV}$ and $C_{IC}$ represent the concentrations of CMV DNA and of IC DNA; $Y_{CMV}$ and $Y_{IC}$ are y-axis intercepts.

In a second step, the amplifications of the CMV DNA in test samples were normalized. To do so, it was assumed that PCR inhibition negatively affects the IC amplification and the CMV amplification equipotent, leading to proportionally diminished amplifications (9). Provided a calibrator sample, which is not inhibited per se, and a test sample contain equal amounts of IC DNA, the IC amplification in the calibrator sample ($CP_{IC-cal}$) can be compared with the amplification of the IC in the test sample ($CP_{IC-sample}$). Inhibition in the test sample can be expressed as $CP_{IC-sample} - CP_{IC-cal}$, and, subsequently, normalized for the CMV-specific amplification by calculating $CP_{CMV} - (CP_{IC-sample} - CP_{IC-cal})$. Furthermore, differing amplification efficiencies of the CMV DNA amplification (Eq. 1) and the IC DNA amplification (Eq. 2) can be standardized by applying the ratio $S_{CMV}/S_{IC}$ to the term $CP_{IC-sample} - CP_{IC-cal}$. Thus, CMV-specific CP values were normalized and standardized by

$$\mathbf{CP_{CMV\text{-}corr}} = CP_{CMV} - [(CP_{IC\text{-}sample} - CP_{IC\text{-}cal}) \times S_{CMV}/S_{IC}] \tag{3}$$

$CP_{CMV-corr}$ represents the normalized CMV-specific CP value of the test sample; $CP_{CMV}$ represents the CP value of the CMV-specific amplification as assessed with the LightCycler instrument; $CP_{IC-sample}$ is the CP value of the IC amplification of the test sample, and $CP_{IC-cal}$ represents the CP value of the IC amplification of the calibrator sample.

**Quantification**

After normalization the concentration of CMV DNA copies in test samples was calculated according to the following equation:

$$CP_{CMV\text{-}cal} - S_{CMV} \times \log C_{cal} = CP_{CMV\text{-}corr} - S_{CMV} \times \log C_{sample} \tag{4}$$

$CP_{CMV-cal}$ corresponds to the CP value of the CMV-specific amplification in the calibrator sample obtained with $C_{cal}$, which is the concentration of CMV DNA in the calibrator sample. This can be solved to log $C_{sample}$ yielding the following:

$$\log C_{sample} = \log C_{cal} - [(CP_{CMV-cal} - CP_{CMV-corr}) / (S_{CMV})] \tag{5}$$

CP values and slope values were exported from the LightCycler's software and imported into Microsoft Excel for spreadsheet calculation of normalization and quantification.

## Results

In a first step, the dynamic range of the competitive PCR was established. When CMV DNA in the range of 10 to 1000 copies per capillary was amplified in the presence of 100 copies of IC DNA both amplifications appeared unaffected from each other. When more than 1000 CMV DNA copies were amplified in the presence of 100 copies of IC DNA, the amplification of the IC DNA was gradually inhibited in competitive fashion (Fig. 2).

Implementation of the CMV-Specific NQC-LC-PCR

In a second step, the slope values representing the amplification efficiencies for the CMV and the IC amplifications were determined. This yielded to mean slope values of –3.537 (SD 0.063, n=5) and of –3.864 (SD 0.068, n=5) for the CMV- and IC-specific amplification, which were subsequently used for the normalization calculations (Eq. 3).

In a third step the normalized quantification of CMV DNA by NQC-LC-PCR was compared with the amount of the CMV reference DNA applied to the PCR and was tested against conventional real-time quantification using graded amounts of the CMV reference DNA as external standard. Quantification with NQC-LC-PCR was found to be strongly correlated with the DNA copy numbers applied to the amplification reaction (Fig. 3a) (r = 0.972; $P < 0.001$; 95% confidence interval [CI], 0.946 – 0.985) and with real-time quantification by external standards (r = 0.971; $P < 0.001$; 95% CI, 0.945 – 0.984) (Fig. 3b).

The limit of normalization of NQC-LC-PCR was examined by applying hemoglobin as the PCR inhibitor. Blood was lyzed and spun down. Free hemoglobin in the supernatant was measured by a standard routine laboratory method and applied to PCR in graded amounts (0.02 µg/ml to 20 µg/ml final concentrations). PCR was carried out by applying 100 copies of CMV reference DNA and 100 copies of IC DNA. Quantification was performed with and without normalized quantification. Hemoglobin inhibited the PCR dose-dependently above 0.02 µg/ml. Complete inhibition was obtained above 2 µg/ml (Fig. 4). When the normalization procedure was applied an almost complete reversion of the inhibition was obtained up to 0.2 µg/ml hemoglobin. At this concentration the PCR was inhibited by 73% (Fig. 4). Hemoglobin concentrations above 0.2 µg/ml resulted to a dose dependent loss of the normalization capability.

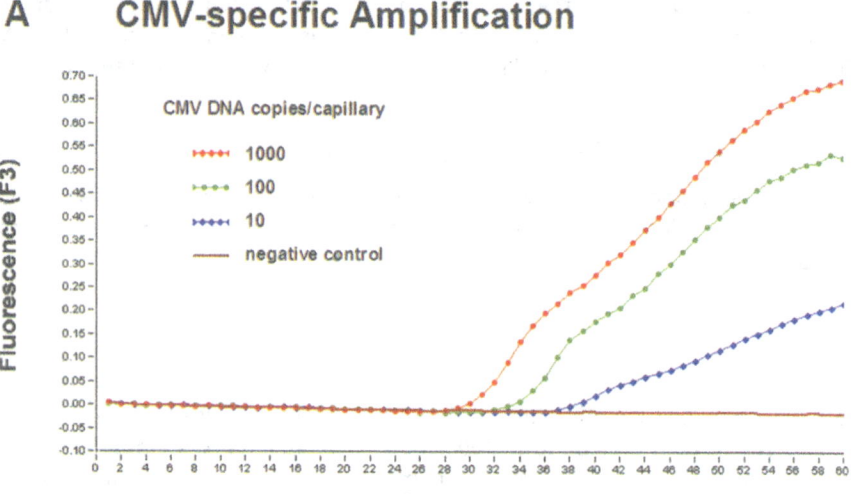

**Fig. 2.** Real-time plot of the CMV-specific competitive LightCycler PCR. CMV DNA in graded amounts was amplified in the presence of about 100 copies of IC DNA. (**A**) shows the CMV-specific amplifications, (**B**) shows the IC-specific amplifications. Note, that the IC amplification was competitively inhibited with increasing amounts of CMV DNA

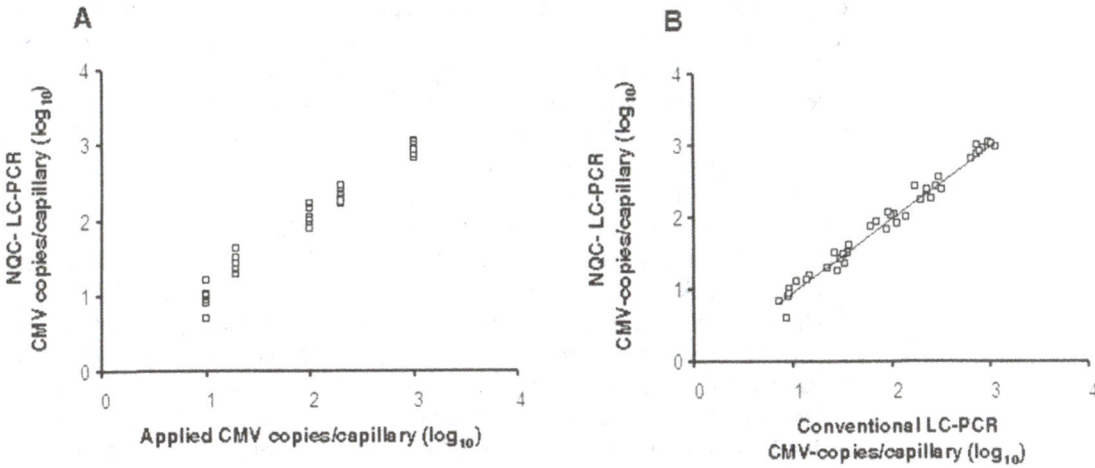

**Fig. 3.** Comparisons of normalized quantification of CMV reference DNA with (**A**) numbers of CMV DNA copies administered to each capillary (10, 20, 100, 200, 1000) (n = 8) and (**B**) with conventional quantification by external standards. Note, that none of the samples was inhibited

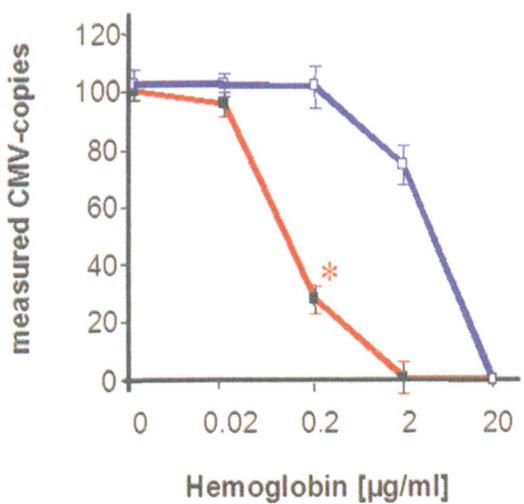

**Fig. 4.** Comparison of normalized quantification of CMV reference DNA with conventional quantification by external standards in PCR inhibited samples. Hemoglobin was added in graded amounts to PCR mixtures containing approximately 100 copies of CMV reference DNA and 100 copies of IC DNA. Quantification was performed in conventional fashion (red curve, closed symbols) and in normalized fashion (blue curve, open symbols). The asterix indicates that the PCR was inhibited by 73 % according to conventional quantification

Quantitative
CMV Assay with
NQC-LC-PCR
For the NQC-LC-PCR based CMV assay calculated 5000 IC DNA copies per ml plasma sample were used. This permitted normalized quantification of CMV DNA from 500 copies/ml to 50 000 copies/ml. When less than 500 CMV DNA copies per ml were used, 250 copies/ml of CMV DNA were detectable. However, due to increased standard deviations reproducible normalized quantification could not be performed (data not shown).

About 80 patient plasma samples and 10 non-infectious plasma samples were examined with the CMV-specific NQC-LC-PCR assay. The IC DNA was found amplified in all samples. The 10 non-infectious samples tested negative for CMV DNA. About 31 patient samples were found positive for CMV DNA, of which 24 samples were found within the range for normalized quantification (Fig. 5). In three samples fewer than 500 CMV DNA copies/ml were detected. In four samples the range for normalized quantification was exceeded, and normalized quantification could not be performed due to competitive inhibition of the IC DNA amplification. Therefore, these samples were diluted and re-tested (Fig. 5).

The normalized quantification of the positive clinical samples was compared with quantification by conventional real-time quantification by testing CMV reference DNA serially diluted in parallel. Both quantification methods were found to be in good agreement and were significantly correlated with each other ($r = 0.973$; $P < 0.001$; 95% CI, 0.943 – 0.987) (Fig 5). None of these samples were PCR-inhibited.

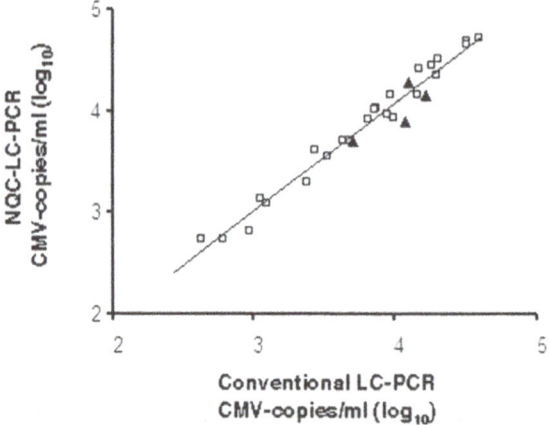

**Fig. 5.** Comparison of normalized quantification of CMV DNA from clinical samples with conventional quantification by external standards. The scatter diagram and regression line shows the relation for 28 positive clinical samples measured with the CMV-specific NQC-LC-PCR assay and with conventional quantitative LightCycler PCR assay using CMV reference DNA in graded amounts as external standard. Four samples exceeded the range for normalized quantification. After a 1:10 dilution these samples were re-tested (triangles). Note that none of the samples was inhibited

# Comments

Accurate PCR-based nucleic acid quantification requires an evaluation of the quality of samples. Therefore, internal controls have been introduced to PCR that are amplified within one reaction vessel either with a different pair of primers or with the same pair of primers as the specific target DNA (3, 4, 10, 11). A prerequisite for this approach in real-time PCR relates to independent detection of IC PCR products and of target specific PCR products. This is met in rapid-cycle real-time PCR, as two different fluorochromes e.g. LC-Red640 and LC-Red705 can be used as label for FRET hybridization probes. Their signals can independently be detected on channels F2 and F3 of the LightCycler's photometer (12). Also, enhanced discrimination between target-specific amplification signals and IC-specific amplification signals can be achieved, when a color-compensating file has been created and applied to the real-time PCR.

It has been suggested that accurate PCR based quantification can only be achieved, when both the internal control specific amplification and the target specific amplification exhibit equal efficiencies. This is thought to rather apply to competitive PCR approaches that amplify both internal control and target specific DNA with a single pair of primers than to multiplex PCR using two pairs of primers (3). With the utilization of a single pair of primers in our approach the efficiencies for CMV specific amplification and internal control specific amplification were found to be not quite equal, however, constant over the dynamic range of the competitive PCR and constant from run to run. Therefore, the amplification efficiency differences could be accounted for with the introduction of the ratio of the two amplification efficiencies into equation 3. Furthermore, the above-mentioned quality of sample was controlled for by using the sample's internal control tested against the internal control of the calibrator sample. Thus, possible amplification variations between test sample and calibrator sample could be normalized by applying equation 3. The potential of this normalization procedure is illustrated by the presented PCR inhibiting experiment. This showed that the normalization procedure in NQC-LC-PCR was capable to compensate for inhibited amplifications by up to about 70% inhibition (Fig. 4) (6). Similar effects were observed when EDTA was used as a PCR inhibitor.

After normalization, quantification by NQC-LC-PCR was performed against the CMV DNA present in the calibrator sample used as sole standard. The comparisons with applied CMV-DNA copy numbers and with the conventional quantification procedure using an external CMV standard showed that the NQC-LC-PCR yields to equivalent quantification results. However, this is only the case when the PCR in conventional quantification is not inhibited. Our findings also suggest that the use of a single CMV DNA concentration complemented with internal control DNA as calibrator sample sufficed for accurate quantification within the defined range of the competitive PCR.

To test the potential utilization of normalized quantification, the CMV-specific NQC-LC-PCR was complemented with automated DNA purification on the MagNA Pure LC instrument. However, manual DNA purification procedures can also be used for NQC-LC-PCR.

The quantification range with the CMV-specific NQC-LC-PCR was set up to meet diagnostic requirements for sensitive detection and clinical relevant quantification of CMV-DNA in plasma samples (10, 13, 14). A range from 500 to 50 000 CMV-DNA copies/ml was chosen. Clinical samples that exceeded the upper limit of quantification were diluted and testing was repeated. This procedure was followed in 4 of the 31 positive samples during our clinical testing. But the range of normalized quantification can be adjusted to cover higher CMV DNA concentrations by increasing the amount of IC DNA added to the samples. However, shifting the range of normalized quantification towards higher CMV DNA concentrations increases the lower limit for normalized quantification. So, as the range for normalized competitive quantification covers only approximately 3 orders of magnitude, NQC-LC-PCR assays need to be adjusted to cover the various diagnostic or analytical needs.

The detection limit of the CMV-specific NQC-LC-PCR assay was found at 250 CMV copies/ml plasma, which was lower than the lower limit for normalized quantification. Due to increased standard deviations reproducible normalized quantification was not possible below 500 CMV DNA copies/ml plasma.

PCR-inhibited samples were not obtained during the clinical evaluation of the CMV-specific NQC-LC-PCR assay, which most likely has resulted to the good agreement of the normalized quantification results with those of the conventional real-time quantification. Several reasons may account for this observation: (i) the selected DNA purification procedure may have removed possible PCR inhibitors yielding to good quality DNA, only; (ii) the sole use of plasma samples for the CMV-assay; (iii) the limited number of CMV-positives samples among the tested clinical samples. Nevertheless, the PCR-inhibiting experiments showed that the NQC-LC-PCR could normalize for inhibited amplifications over a wide range.

Apart from normalized quantification the utilization of the IC in test samples permitted control of the CMV-specific NQC-LC-PCR assay from DNA purification through the amplification process. Thus, PCR-inhibited samples should be detectable with the described methodology.

In conclusion, the described methodology for normalized quantitative competitive real-time PCR on the LightCycler instrument should prove useful for the accurate assessment of other microbial targets.

## References

1. Mackay IM, Arden KE, Nitsche A (2002) Real-time PCR in virology. Nucleic Acids Res. 30: 1292–1305
2. Al-Soud WA, Radström P (2001) Purification and characterization of PCR-inhibitory components in blood cells. J. Clin. Microbiol. 39: 485–493
3. Bai X, Rogers BB, Harkins PC, Sommerauer J, Squires R, Rotondo K, Quan A, Dawson DB, Scheuermann RH (2000) Predictive value of quantification PCR-based viral burden analysis for eight human herpesviruses in pediatric solid organ transplant patients. J. Mol. Diagn. 2: 191–201
4. DiDomenico N, Link H, Knobel R, Caratsch T, Weschler W, Loewy ZG, Rosenstraus M (1996) COBAS AMPLICOR: fully automated RNA and DNA amplification and detection system for routine diagnostic PCR. Clin. Chem. 42:1915–23

5. Sia IG, Wilson JA, Espy MJ, Paya CV, Smith TF (2000) Evaluation of the COBAS AMPLICOR CMV MONITOR test for detection of viral DNA in specimens taken from patients after liver transplantation. J. Clin. Microbiol. 38:600–606

6. Stöcher M, Berg, J (2002) Normalized quantification of human cytomegalievirus DNA by competitive real-time PCR on the LightCycler instrument. J. Clin. Microbiol. 40: 4547–4553

7. Stöcher M, Leb V, Bozic M, Kessler HH, Halwachs-Baumann G, Landt O, Stekel, H, Berg J (2003) Parallel detection of five human herpes virus DNAs by a set of real-time polymerase chain reactions in a single run. J. Clin. Virol. 26: 85–93

8. Stöcher M, Leb V, Hölzl G, Berg J (2002) A simple approach to the generation of competitive internal controls for real-time PCR assays on the LightCycler. J. Clin. Virol. 25 Suppl 3:45–53.

9. Siebert PD, Larrick JW (1992) Competitive PCR. Nature 359:557–558

10. Boeckh M, Boivin G (1998) Quantitation of cytomegalovirus: methodologic aspects and clinical applications. Clin. Microbiol. Rev. 11: 533–554

11. Lyon E, Millson A, Lowery MC, Woods R, Wittwer CT (2001) Quantification of HER2/neu gene amplification by competitive PCR using fluorescent melting curve analysis. Clin. Chem. 47: 844–851

12. Elenitoba-Johnson KS, Bohling SD, Wittwer CT, King TC (2001) Multiplex PCR by multi-colour fluorimetry and fluorescence melting curve analysis. Nat Med 7:249–253

13. Halwachs-Baumann G, Wilders-Truschnig M, Enzinger G, Eibl M, Linkesch W, Dornbusch H.J, Santner BI, Marth E, Kessler HH (2001) Cytomegalievirus diagnosis in renal and bone marrow transplant recipients: the impact of molecular assays. J. Clin. Virol. 20:49–57

14. Sia IG, Wilson JA, Espy MJ, Paya CV, Smith TF (2000) Evaluation of the COBAS AMPLICOR CMV MONITOR test for detection of viral DNA in specimens taken from patients after liver transplantation. J. Clin. Microbiol. 38:600–606